Environmental Consciousness in China

Environmental Consciousness in China

Change With Social Transformation

QIU ZHONG

GUOQING SHI

ELSEVIER

CP
CHANDOS PUBLISHING
An imprint of Elsevier

Chandos Publishing is an imprint of Elsevier
50 Hampshire Street, 5th Floor, Cambridge, MA 02139, United States
The Boulevard, Langford Lane, Kidlington, OX5 1GB, United Kingdom

British Library Cataloguing-in-Publication Data
A catalogue record for this book is available from the British Library

Library of Congress Cataloging-in-Publication Data
A catalog record for this book is available from the Library of Congress

ISBN: 978-0-08-100388-6

For information on all Chandos Publishing publications
visit our website at https://www.elsevier.com/books-and-journals

Publisher: Glyn Jones
Editorial Project Manager: Rachel Pomery
Production Project Manager:
Surya Narayanan Jayachandran
Cover Designer: Matthew Limbert

Typeset by MPS Limited, Chennai, India

Contents

Preface

Environmental problems are essential problems of human beings. Over the past years, environmental problems became more serious and more multiplicity, which is the fundamental reason for the concern on the environment. The environmental problems of China are consistent with the world's environmental problems, which are the secondary environmental problems induced by human activities. In recent days, the confliction between the increase in economics and the deterioration of the environment in China becomes crucial. In addition, the aggravation of the contradiction between environmental problems and social developments makes the consciousness of the environment more important. Moreover, the secondary environmental problems induced by human activities bring the focus of the contradiction from the objective environment to the subjective human activities. Therefore, proper guidance, influence, and change in the personal activities to satisfy the environmental protection requirements are the vital factors of the success of the environmental protection project. It is also the core of the realization of sustainable development in China.

At the present time, the explosive stage of the environmental problems overlaps with the promoting period of the urbanization process in China. The systematic research of the environmental consciousness and activities at this crucial transition time is very important not only for the promotion of people's environmental consciousness, proper guidance of the people's environmental activities, but also to realize the goal of sustainable development in the future. Firstly, as a social phenomenon, environmental consciousness is both an important area of social science research and one of the attentive focuses of the government, the media, and the public. Secondly, environmental activities keep changing with both the subjective environmental deterioration and the increase of the environmental consciousness. However, the relationship between environmental activities and environmental consciousness is still under further research. The environmental consciousness reveals the guiding role in the behavior of both environmental protection and the environment struggles. Therefore, further research on the environmental consciousness and more comprehensive exploration of the relationship between environmental consciousness and environmental activities will have important science and application values.

The research in this book significantly differs from the traditional environmental consciousness research in the following aspects: the most significant innovation in this book is including the urbanization process into the discussion, which is the projection of the real situation in China. In the research methods and the research designs, we compare the environmental consciousness and activities between the urban residents, the rural residents, and the immigrate residents. This will be deeper research on the mechanics of environmental consciousness and activities. After that, the discussion of the influence factor on environmental consciousness is one of the main points in this book. It is also the first research that compares the population statistical factor and the theoretical dimensional factor. Third, in environmental activity research, the three dimensions, that is, the environmental protection willingness, the environmental protection activities, and the environmental struggles, respectively, reflect the shallow layers of the environmental protection tends, the actual environmental protection activities, and the deep layers of the policy interactions or the environmental struggles. Finally, this is the first research that verifies the applicability of the value-belief-norm theory in China, and this is the first time to systemically construct an environmental consciousness-activity model of the residence and the immigration.

This book is constructed as follows: the first chapter reviews several typical environmental disasters and shows the stories that kept repeating. The second chapter shows the current environmental problems both in China and over the world. Chapter 3, From Movements to Measurement, compares the development of environmental consciousness both in China and over the world. The development of environmental consciousness research is also briefly reviewed in this chapter. Chapter 4, From a Longitudinal Study to Cross-Sectional Study, introduces the research design, the research method, and the sampling method used in this book. This chapter discusses the transition of environmental consciousness in China and analyzes the population statistical variables and eras on the macroscale. Chapter 5, What Drives Environmental Consciousness?, mainly explains the mechanics of the transition of environmental consciousness in China. In this chapter, the factors which affect this transition are also analyzed. The analyses in the sixth chapter are focused on environmental behavior in the results of the survey. Chapter 7, The Path to Environmental Behaviors, tries to construct a model to analyze environmental consciousness and environmental behavior. The eighth chapter discusses the environmental consciousness under the tide of the time. The last chapter is a conclusion of the research in this book and a comparison with the research results of other countries and regions.

This book may contribute to environmental sociology, immigration sociology, environmental science, or the management researches or studies. It may also be a reference for the environmental departments of the government. For the researchers studying environmental consciousness in both China and other countries, this book may be a reference too. Welcome and appreciate any discussions.

Acknowledgments

Writing a book is like a journey. It was more difficult than I imagined, but also more real and rewarding than I could have ever pictured before. None of this would have been possible without my family. From reading early drafts to commenting on a single graph, my husband Lichao stood by me in getting this book done. Thank you my love! Moreover, my baby boy was in my womb when I was writing this book. Thinking that he may read those words someday, indeed makes this journey full of joy. Thank you for coming to us at the perfect timing my dear Zachary!

Words on the paper get power when someone reads them. I want to thank all those who have been a part of this book: Glyn Jones; Peter W. Adamson; Rachel Pomery; Jennifer Pierce; Emily Thomson; Narmatha Mohan; Unni Kannan Ramu; Sathya Narayanan; and Surya Narayanan Jayachandran. If any of their efforts were missing you would have not seen this book. Sincerely I thank each one of you!

Qiu Zhong

CHAPTER ONE

Stories repeat again and again

1.1 England, 19th century

Back in December 1873, several severe "fog" swept London within 1 month. Visibility was reduced to few yards. *Time* issued a statement "all locomotion, especially to people on foot at the crossings of great thoroughfares, berceuse extremely dangerous" (Freese, 2004). People were shocked at that time, but they did not know they will continue suffering the weather even worse in the later years. Fig. 1.1 shows a photo taken the heavy smog in Piccadilly Circus, London, on 6th December 1952.

The death statistics were released several decades later. About 270–700 London people were silently killed by the 1873's "fog." The death number was between 700 and 1100 when London suffered her 1880's "fog" week. Death number was 1000 for the 1892's. While these deaths went largely unreported (Freese, 2004).

In the year 1952, no one can forget from Friday December 5 to Tuesday December 9. Even now, "The Great Smog" or "Big Smoke" typically tagged to the 1952s London air pollution. "The Great Smog" not only caused rapid growth of traffic accidents but also directly brought health problems. The government reported a surge of death at this time. Just 4 days later on December 8, medical department gave a report that 4000 people had died prematurely because of the smog (Brimblecombe, 1987). According to a later study, 12,000 deaths were considered related to the smog (Bell & Davis, 2001).

1.2 The United States, 20th century

1.2.1 July 1943

England is not the only place that has smog. Smog also shows up in the United States. Fig. 1.2 shows the first recorded photo of smog in Los

Environmental Consciousness in China.
DOI: https://doi.org/10.1016/B978-0-08-100388-6.00001-5

Figure 1.1 Nelson's Column during the Great Smog. *<https://en.wikipedia.org/wiki/File:Nelson%27s_Column_during_the_Great_Smog_of_1952.jpg>*.

Angeles, California, in 1943. The entire city was suddenly filled with some unknown gas with pretty low visibility. Some thought it might be a World War II attack by Japan (Butcher, 2013). Like the "fog" in London, the smog in Los Angeles, California also caused a fatal increase in traffic accidents and health issues. The story was repeating in the United States.

1.2.2 October 1948

Donora is a small industrial-based city with a steel mill and a zinc works lying in a river valley along the Monongahela River near the city Pittsburgh. A few years after the smog sweeps Los Angeles, in 1948, the city was all in dense yellow smog on October 27, and it lasted for a few more days. Fig. 1.3 shows a photo took in the smog during the daytime on October 30, 1948. The sun was completely blocked and the visibility was very low. During the smog, more than 20 people died and 600 were hospitalized (Tracton, 2012). Around 40% of the total population in the city was affected, concluding the official epidemiological study

Figure 1.2 Smog in Los Angeles, California (UCLA Library, 1948).

(Townsend, 1950). The story repeated one more time and was repeating in one of the largest cities in the United States, New York.

1.2.3 November 1966

From November 23–26 in 1966 was the Thanksgiving holiday weekend that year. Thanksgiving celebrations were all across the United States. Residents in New York City, an extra-large metropolis located in the northeast of the country found dense smog covering the city in their Thanksgiving "gift" as shown in Fig. 1.4. This environmental disaster caused many health issues, for example, stinging eyes, coughing, and respiratory distress. About 168 people died likely due to the smog (Bird, 1967). After having a look at the stories in the Europe and the United States, we saw about Asia.

Figure 1.3 The smog in Donora on October 29, 1948 at noon. *NOAA Ocean service Education.*

1.3 Japan, 20th century

1.3.1 1960 to 1972

Yokkaichi is a Japanese city lying beside the Ise Bay. During the 1950s, many oil industries rose in this city and started to cause air pollution. Severe smog was built in the city. Fig. 1.5 shows a photo with a large amount of air pollution posted on the official website of Yokkaichi city introducing the history at that time. After a few years, fish caught in Ise Bay started to have a bad taste and the residents in Yokkaichi developed asthma. This Yokkaichi asthma then became one of the Four Big Pollution Diseases of Japan.

When these environmental disasters reached the United Kingdom, the United States, and Japan in 19th and 20th centuries, they were both on the stage of the fastest economic growth. China, right now, is at the same stage. The story is repeating.

1.4 China, 21th century

1.4.1 April 15, 2015

The photo shown in Fig. 1.6 was taken at Tiananmen Square on April 15, 2015. Beijing was hit by the huge sandstorm from Wednesday evening to

Figure 1.4 A photo published on the front page of the New York Times, showing the city in smog. Courtesy of Andy Blair.

that night. A yellow alert was issued in the afternoon by the Meteorological Observatory, that the visibility would drop below 1000 m (Miou, 2015). This sandstorm was considered the worst one over the past 13 years. Authorities urged people to stay indoors. "The apocalyptic metaphor" was used to describe the sense by many internet new (Yang, 2015).

Beijing Municipal Environmental Monitoring Center reported that the PM10 air pollution value reached 1000 $\mu g/m^3$ on that day. This level was considered hazardous (Beijing Environmental Monitoring Center, 2018). At 7:00 p.m. on that day, PM2.5 particulates value appeared as 722 in Beijing, according to the U.S. Department of State Air Quality Monitoring Program (U.S. Department of State Air Quality Monitoring Program, 2018). Fig. 1.7 shows the pollution made the Tiananmen Square hardly seen, on December 1st, 2015.

1.4.2 December 1, 2015

Beijing Municipal Environmental Protection Monitoring Center showed data from the southwest area of Beijing at 6:00 p.m. on November 30.

Figure 1.5 Yokkaichi asthma. <*https://www.monodukuri.com/jirei/article/499*>.

The PM2.5 concentrations were close to 1000 μg/m^3. At 8:00 a.m. on the next day, the PM2.5 particulates value for the past 24 hours appeared as 745 μg/m^3 (Kong, 2015).

Value exceeding 300 will be considered hazardous and dangerous to human health. BBC Chinese published news on that day titled "Beijing suffering severe haze, people online saying 'Beijing disappeared'" (Liu, 2015). The extremely poor quality of the air prompted the city officials to issue the first-ever red alert on December 8, 2015 (Sui-Lee Wee, 2015).

These 2 days mentioned above are not the only days with heavy air pollutions; however, the historical data of PM2.5 shows more than 29 days having a PM2.5 value more than 500 from 2014 to the mid of 2017, as shown in Fig. 1.8. In the figure, the days with more PM2.5 happened every winter within the plotted range, and the winter of 2015 and 2016 had more days of higher PM2.5. So the stories are still repeating.

Does nature fog get this huge power that could make so much damage, or sometimes killing so many people? Of cause not. People call it "fog" because it visually likes the fog and no one considered it an environmental disaster at that time. When similar situations repeated in the United States, and China many decades later than it shows up in the United Kingdom, people still did not connect "fog" with environmental issues initially. The United States thought it might be a World War II attack by Japan when

Figure 1.6 A photo shows the Tiananmen Square in sandstorm (Lan, 2015).

Los Angeles suffered its first unexplainable, throat-burning noxious gas attack in 1943 (Butcher, 2013). Similarly, in China, when people started to question why they suddenly got daunting air pollution without any sign latterly, the answer is that it was called "fog" (Chai, 2015).

When the truth been revealed, people finally realized those environmental disasters were caused by human beings. This not only includes environmental disasters related to air pollutions, but also included many other environmental disasters such as water, soil, biological, radioactive contamination, noise, thermal, and electromagnetic interference pollutions that are also caused by human beings. In this case, US's suspect was the closest one; however, they found that the disaster was not cast by their enemy, but by their selves.

1.5 Economic development echo?

Nowadays, China's environmental issues have been highlighted in the mass media. They cannot be ignored, but we need to keep in mind that those environmental issues were not just appeared in China. The typical examples are England and America encountered the similar "smog" situation and other environmental deteriorations before. More interesting things for the researchers are the logic behind those repeated stories.

Figure 1.7 The second red forecast on Dec. 21, 2015. <*https://zh.wikipedia.org/wiki/2015%E5%B9%B4%E8%87%B32016%E5%B9%B4%E5%86%AC%E4%B8%AD%E5%9B%BD%E5%8C%97%E4%BA%AC%E9%9B%BE%E9%9C%BE%E4%BA%8B%E4%BB%B6*>.

Although the names in these stories are changed from "fog," "smog" to "PM2.5," the basic chemistry did not change too much. Coal takes the main responsibility for them, but coal also is the foundation stone for their social and economic development.

To better understand the relationship between social and economic development and the environmental disaster, let us take a look at a brief review of the historical economic development of the aforementioned four countries.

From the 1750s to the 1830s, the industrial revolution has occurred first in England. Innovations with steam power and coal impelled Great Britain to global economic dominance. Fig. 1.9 shows during this time, the gross domestic product per capita or the GDP per capita in the United Kingdom was greatly increased. The United States in the 20th century followed up the second phase of the industrial revolution from 1870 to 1900. Innovations, for example, electricity, the internal combustion engine, and oil/gas refining, boosted the economy rise in the United States. This is also a period that the GDP of the United States rapidly increased, as shown in Fig. 1.9. These make Great Britain and the United States the leading countries in economics in the world. As for Japan, after World War II, it was under several time periods with fast economic

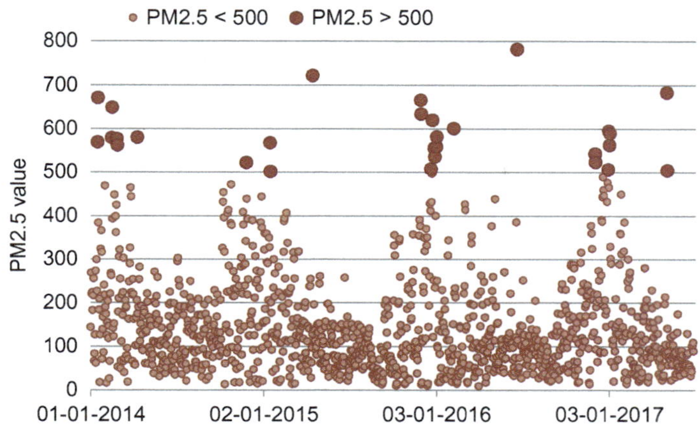

Figure 1.8 PM2.5 historical data of Beijing. *From U.S. Department of State Air Quality Monitoring Program. 2018* Beijing air quality historical data. *<http://www.stateair.net/web/historical/1/1.html>.*

growth, that is, the Jinmu Boom, the Iwato boom, the economic boom created by the Tokyo Olympic Games, and the Izanagi boom. During these economic booms, the Japanese government created many manufacturers in producing TVs, washing machines, cars, and also steels. These were later called the Japanese economic miracle, and the GDP during this time was almost doubled. From here, we can see the repeated stories in the last few sections are all happened during a rapid economic boom.

At the beginning of the 21st century, when the aforementioned stories in the last section happened in China, China emerged as the world's largest manufacturer, largest holder of foreign exchange reserves, largest merchandise trading economy, the second-largest destination of foreign direct investment (FDI), and is projected to become the largest economy in the world in 2014 (Morrison, 2013).

China reached its highest economic level in the fastest economic growth rate which was never seen. There are three approaches reflected in China's huge economic development. First of all, GDP in China annually increased by an average of 9.8% after reform and opening since 1978. At the same time, the nominal GDP per capita increased by 50 times and more than 700 million people shake off poverty in China (Xinhua Daily Telecommunications, 2016).

Second, most people became wealthier. A report from the World Bank pointed out that from 1981 to 2004, the poverty population

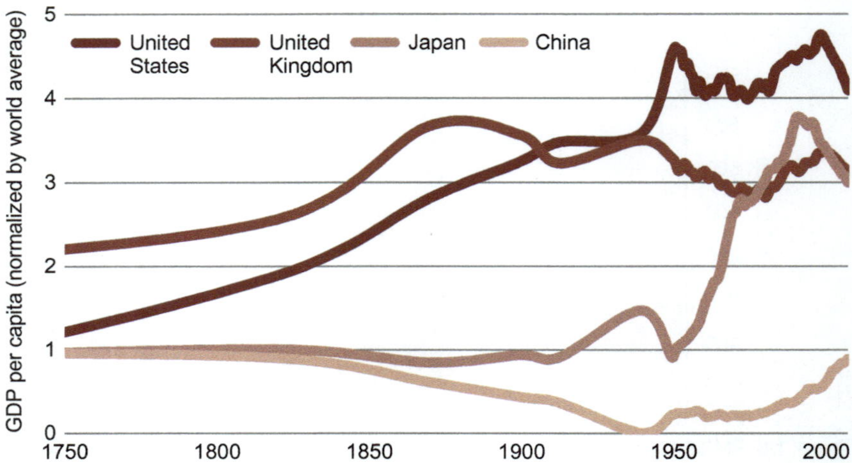

Figure 1.9 Historical trends in GDP per capita in the United States, the United Kingdom, Japan, and China. *From Maddison, A. 2008.* Angus Maddison's homepage. *<http://www.ggdc.net/maddison/oriindex.htm>*.

decreased from 652 million to 135 million (World Bank, 2009), which means the poverty population ratio dropped significantly from 65.2% to 10.4% (National Bureau of Statistics of China, 2000–2018). Here, the absolute poverty standard at 2003 rural prices, that is, consumption of a dollar per day per capita is used (World Bank, 2009). A story can be told by mobile phones: in China, the first cell phone for sale was in 1987. At that time, there were three threes for the cell phones: a cell phone is 30,000 Chinese yuan (around $3600), weight 3 pounds, and you need to wait 3 months to get the phone after you have paid. Travelling to the year 2012, 82 people in every 100 have at least one cell phone, in other words, almost everyone in China has one (Wei, 2013).

Third, China has much more foreign exchange reserves. In 1978, China only has held $167 million foreign exchange reserves and ranked 38th worldwide. After the rapid economic development and growth, in 2006, China surpassed Japan and ranked first in the world, with more than 1 trillion US dollars in foreign exchange reserves (Wei, 2013).

Overall, the United Kingdom, the United States, and China all suffered serious environmental problems after they became one of the world's biggest economic bodies. Stories repeat again and again. It seems like the development echo.

CHAPTER TWO

Environmental issues to protection: the United States versus China

The environmental issues are the harmful environmental phenomena on human survival and development in a region of over the world. Here, harmful environmental phenomena may be due to the human activities or not. Based on this, environmental issues may be classified into two types. The first type is "the native environmental issues." This type of issue is induced by natural evolutions or natural disasters, for example, earthquakes, floods, droughts, typhoons, avalanches, landslides, and mudslides. The other type is human activities induced environmental issues and this type is named as "the secondary environmental issues" or "environmental hazard." Fig. 2.1 shows an illustration of the classification. Nowadays, environmental issues are often referred to the latter case. One broadly noticed definition in the modern literature is: "An environmental problem is any change of state in the physical environment which is brought about by human interference with the physical environment and has effects

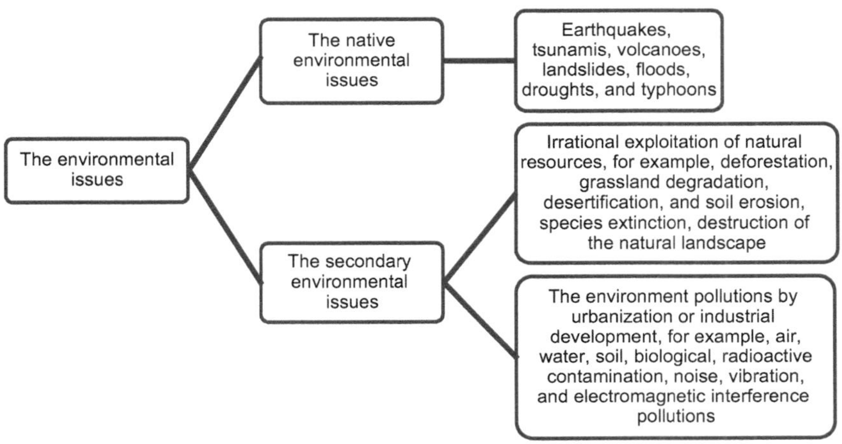

Figure 2.1 The classification of the environmental issues.

Environmental Consciousness in China.
DOI: https://doi.org/10.1016/B978-0-08-100388-6.00002-7

which society deems unacceptable in the light of its shared norms" (Blowers, 1995). Based on the definition, the secondary environmental issues can be divided into two categories: the damage caused by the exploitation of natural resources and the environmental pollution caused by industrial development or urbanization, as shown in Fig. 2.1.

2.1 Ancient era—environmental issues emerges

In the agricultural civilization, environmental issues started to emerge in both western countries and China. As the population increased, agriculture increased; there was more and more interaction between humans and the environment. People gathered together and large cities were formed. Better tools, increasing the number of livestock, and more resources usage brought environmental issues. In this stage, most issues are the secondary environmental issues under first category, that is, the damage caused by the exploitation of natural resources.

Hughes (2014) discussed some typical environmental issues of ancient Greece and Rome, that is, soil erosion, wildlife depletion, industrial pollutions, agricultural decrease, and urbanization issues. In the United States, numerous cultures of indigenous people were formed before the 17th century; however, the record of that time is very limited.

In China, it is believed that environmental issues also started. Chang'an, one of the largest cities in ancient China, has been the capital of many dynasties. The underground water in this ancient city is believed to be polluted starting from the Han dynasty in CE 582 (Chou, Huang, & Zhou, 2007). The pollution was increasing, and in Tang to Song dynasty from around CE 700–1000, the taste of underground water was salty and bitter. The water was almost not drinkable due to the very bad taste in the Qing dynasty in around CE 1800. The pollution of the underground water of this city is believed due to the increase in the population and thus the increase of the poorly treated waste.

Another example shows in Guanzhong Plain, a historical region in the middle region of China surrounded by Loess Plateau, Qinling mountains, and Daba Mountains. Four passes enclose this densely populated and rich plain and makes it a place contested by all strategists. A large population in the plain causing a large amount of usage of lumber and reclamation of lots of lands, which both largely reduced the forest area of the surrounding areas. For example, historical data shows in the Zhou dynasty from around 1000 to 200 BCE, the forest area of the Loess Plateau near Granzhong Plain was 79 million acres. The area reduced to 62 million acres in the Qin and Han dynasty from around 200 BCE to CE 200 and further reduced to 49 million acres in the Tang and Song dynasty from around CE 600–1200. In the Ming and Qing dynasty from around CE 1400–1900, the forest area was only 20 million acres (He & Lang, 2009). Due to the artificial land reclamation in ancient China, the area of the forest was largely decreased and thus caused soil erosion. The soil erosion damages the rest vegetation and caused further soil erosion.

These examples show that environmental issues existed in the ancient era. However, during that era, the issues were not being highlighted, and very rare emperors deeply thought about it. Even if they notice the problems, moving and polluting again will be their solution. In this era, the environmental issues were usually not severe, and there is almost no environmental consciousness during this era.

2.2 1950s to 1960s—awakening in the United States

Western society not only has the lesson and summary about the relationship with the environment, but also the waked up environmentalists became a forerunner in the western society, while developing countries just have their industrial civilization and did not arouse any consciousness until later.

2.2.1 The United States: budding and awakening

At the end of the 19th century, the United States completed the process of modern industrialization. The total industrial output value began to exceed the total agricultural output value. At the end of the 1920s, the realization of electrification began. In the 1930s, the United States suffered a serious economic recession, along with natural disasters. Out of this crisis, President Franklin Roosevelt took a number of measures, one of which was resource protection legislation, providing work as a form of relief, such as planting trees and grass, and building irrigation. The most outstanding effective measures were the comprehensive treatment of the Tennessee River Valley, the Great Plains dust governance of the United

States, and the building of a series of water conservancy and hydropower projects in the western area.

In the 1950s, the invention of the transistor caused the United States another scientific and technological revolution. Keypad telephones, stereo radios, drying/washing machines, air conditioners, and frozen foods emerged at that time. The US economy had experienced unprecedented prosperity. As the Americans enjoyed wealth, a blemish of the industrial society appeared gradually. Water pollution, nuclear pollution, chemical pollution, and air and noise pollution caused by the automotive industry, followed the rapid economic development.

The water pollution control law, which was promulgated in 1948, marked the beginning of the federal government's involvement in environmental protection. However, the environmental pollution of the United States did not become a global issue at that time. The lack of the financial capacity of the federal government limited environmental protection in this period as well. Local governments took care of most of the management responsibilities and focused on the protection of water resources. In conclusion, the effects in different areas varied a lot.

In the 1960s, environmental deterioration, as well as environmental consciousness, was generated among the elite of the United States. In 1962, "Silent Spring" written by American marine biologist Rachel Carson was published with a large number of facts. She demonstrated that industrial pollution damages the life on the earth including human beings and stated that the industrial revolution caused ecological damage and first alarmed the world to the seriousness of environmental protection. Carson said in the book: "For the first time in the history of the world, every human being is now subjected to contact with dangerous chemicals, from the moment of conception until death" (Carson, 1962). The book was widely read and made the public concerned about contamination. In 1972, Carson's work facilitated the ban on using DDT in the United States.

2.2.2 China: no consciousness

When the new China was just founded, the Chinese industrial foundation was very weak, the population was not too large, and the environment was not a problem. At that time, environmental protection was only considered in keeping the work and living place clean and improving the outlook of urban and rural areas. These were the main contents of the patriotic public cleaning campaign. This situation continued until the mid-1950s. Due to the limited scale of production, environmental problems had no major effect on economic development. From the country's overall situation, the main environmental issue was the regional ecological damage and contamination. There was no true sense of the concept of environmental protection in this period; neither was progressing toward the development of related laws and regulations (Qu, 2000).

After the 1950s, the environmental pollution of China became more and more serious. Antipollution campaigns appeared in some developed countries. In China, people had some preliminary understanding of environmental protection. However, most viewed environmental protection as only related to air and water pollution control, solid waste treatment (called "three wastes" treatment), and the elimination of noise with technical measures and management. The main purpose was to eliminate pollution to protect human health.

In addition, since 1957, China experienced the political situation (Great Leap Forward, Great Steel-Iron Making, the People's Commune Movement), and the economic situation (3 years of natural disasters and food shortages), which made people strongly demand natural resources. However, there was no comprehensive understanding of the proper use of natural resources. Most people were neither familiar with the concept of environmental protection; nor understand the situation in this area. They thought Chinese environmental problems were not important and that they need not be too anxious. At the same time, most people under socialism did not admit there was a problem of environmental pollution. They thought that it was a product of capitalism and was an incurable disease of the capitalist countries. Whoever said that China had a pollution problem was negating the superiority of socialism. Some people thought environmental issues were the same as health problems, overemphasizing the damage to human health by environmental pollution while minimizing the seriousness of environmental pollution's harm to society in all other areas.

In the science area, the book Geographical Environment and Social Development (Wu, 1950) published in China written by Ze Wu discussed the geography of social development within the Marxist perspective. At the same time in the 1950s, there was issued another work about the relationship between the natural environment and human society Geographical environment, population, and social development (Wang, 1955) written by Zhende Wang based on Wu's writing. Wang further added population variables to strengthen the understanding of the relationship between the geographical environment and social development.

2.3 1960s to 1970s—initiation stage in China

As the environmental movement in full swing in the United States from the 1960s, China just noticed the environment may be an issue a decade later. The lessons and experiences that the United States had taken will be valuable for other countries, including China, to learn.

2.3.1 The United States: comprehensive transformation

The study, Longitudinal Studies of Public Concern for Environmental Quality (1968–76), shows that the peak of public concern for environments was at the end of 1969 and the early part of 1970 (Dunlap & Scarce, 1991). The 22 USC Section 2151p-1, adopted by Congress through P.L. 87-195 December 30, 1969, as amended, provides for the protection of tropical forests, by placing a high priority on "conservation and sustainable management of tropical forests" and to "the fullest extent feasible." With this basic environmental law, environmental protection came onto the legalization.

On April 22, 1970, an estimated 20 million people across the nation attended the environmental movement with the goal of a healthy, sustainable environment (Mitchell & Davies, 1978). It was then celebrated as Earth Day and was marked as the birth of the modern environmental movement with the public concern for environmental quality reaching unprecedented levels. This supportive public opinion provides a valuable resource for the environmental movement. The environmental movement has had a great impact on environmental consciousness and environmental protection, even on socioeconomic policy and the future of the movement will depend heavily on the degree to which environmentalists can effectively mobilize the support (Dunlap & Mertig, 1992).

In the 1970s of the last century, the United States faced too many environmental pollution issues such as acid rain, nuclear pollution, dredged material pollution, oil pollution, and so on. Because of these issues, President Nixon officially brought the United States Environmental Protection Agency (EPA) into existence in 1970. His plan was to establish EPA for pollution control purposes, because "the environment must be perceived as a single, interrelated system" (Nixon, 1970). From then on, the US Congress passed many environmental bills tackling everything from air and water pollution and garbage to protection for fisheries, marine mammals, and endangered species.

As environmental issues appeared on the front page of newspapers and the covers of magazines, words and phrases such as ecology, environmental costs, resource depletion, eutrophication of rivers, environmentalism, and other similar words became popular even in several famous magazines, for example, Times, Fortune, Life, The New York Times, Washington Post, and so on.

The Club of Rome was founded in April 1968 and raised considerable public attention in 1972 with its report "The Limits to Growth." It is a book modeling the consequences of a rapidly growing world population and its finite resource supplies. After that, "only one earth" became the slogan of the Stockholm Conference or the United Nations Conference on the Human Environment (Ekins, 1993).

The trend data in Gallup surveys, especially covering the years from 1969 to 1970, shows the change in environmental consciousness. About 17% of respondents selected pollution in 1965, make it ranked ninth among the 10 top problems. In 1970, this number is more than tripled: 53% selecting it and it becomes the second problem only after the crime reduction (Dunlap, 1995).

2.3.2 China: initiation stage

The Cultural Revolution put the Chinese economy onto the verge of a collapse between 1966 and 1976 (Qu, 2002). Environmental problems have become more and more serious. Several environmental issues showed destruction and depredation at that time. In 1972 an emergency caused by pollution happened at Dalian Bay. The rising tide brought black water and ebb tide left the beach black. The pollution left more than 800 acres of dead shellfish on the beach. Every year China loses more than 20,000 pounds of trepang, more than 200,000 pounds of shellfish, and more than 300,000 pounds of corbicula. Some fishermen even acquired Shui Yu (methylmercury poisoning) sickness from eating those fish or shellfish.

With the advice of Premier Enlai Zhou, China sent one delegation to participate in the Stockholm Conference on June 5, 1972. This Human Environmental Conference pointed out that environmental problems are not just linked with technology but also have a close relationship with socioeconomics; these problems not only makes the issues local but also global. This conference is recognized as the beginning of environmental concepts that promoted environmental protection in China. The senior leader of China paid much attention to environmental issues after this conference, and the environment became significant along with social and economic development.

On May 20, 1973, the Chinese Council held the first national environmental protection meeting in Beijing. Protection policy increased in this meeting for the purpose of planning and distributing protection of the environment for the masses of people. On November 17, 1973, three state Commissions co-promulgated the "industrial emissions test standard." This "standard" provided the basic law for the EPA, ended the unsystematic policy of Chinese environmental protection, and also made feasible environmental monitoring.

Dependence upon those policies was influenced by the Chinese society; people began to understand the relationship between environments, and developed and realized that environmental protection was not only to control but also (or more importantly) to utilize resources rationally. Economic growing is limited by environmental capacity.

2.4 1970s to 1980s—laws established

2.4.1 The United States: the recoil after the recession

In the early 1970s, after the peak of the environmental movement, early environmental protection policies made progress. As a result of sufficient funds and great attention from many people in the United States to the pollution problems, environmental protection methods were greatly enhanced. This enhancement further improved the coordination capacity of the environmental protection systems on all levels. However, the huge expense of environmental protection in the United States became an important component of budget deficits. Strict control measures also caused a negative effect on the economy and this caused great concern for both the US government and the public.

During the years 1973–75, the United States suffered a structural crisis which was characterized by economic stagflation. At that period, the United States and the Western world had a serious economic crisis. The panic over a shortage of resources (especially oil) made environmental concerns take backstage in the United States. The economic issues had a more profound impact on people's lives than environmental issues. In 1984, compared with 1980 before Ronald Reagan was elected, the laws and regulations published by the Government to intervene in social activities were reduced a lot, greatly easing the standards and regulations in environmental protection, technology security, health care, and consumer protection which published previously.

Nevertheless, after a brief setback in the 1970s and 1980s, the environmental concerns in the United States resumed. Time Magazine's cover named Endangered Earth the planet of the year was found this quote "For man has reached a point in his evolution where he has the power to affect, for better or worse, the present and future state of the planet" (Sancton, 1989).

Unlike the flourishing environmental movement in the early 1970s, at this time environmental consciousness was deeply ingrained in American people's hearts.

2.4.2 China: keep up with the pace

China established environmental protection as a basic national policy and specified a three policy system: the first is integrating prevention with control. Second, the-polluters-pay tightening up of environmental management to include payment of pollution cleanup. The third is reaping benefits from environmental protection (Zhou, 2009).

In 1978, China established the National Committee for Man and Biosphere. In 1980, China and the United States signed documents on cooperation in environmental protection, which initiated the process of China's bilateral cooperation in environmental protection. In 1983, China established the first national-level professional newspaper on environmental protection—"China Environment News." The second national environmental protection conference convened in 1983, emphasized environmental protection as a basic national policy of China. This strategic policy of environmental protection of China included: synchronized planning, implementation, and development in terms of economic and urban and rural development and environmental protection, bringing about a harmony of economic returns and contribution to society and environmental protection. On September 11, 1989, China signed the "Vienna Convention for the Protection of the Ozone Layer," which was enforced in China on December 10 of the same year. Then China also joined the "Montreal Protocol on Substance" in 1991 and continues to raise people's consensus on the protection of the Ozone Layer. The Environmental Protection Law of the People's Republic of China adopted at the 11th Meeting of the Standing Committee of the Seventh National People's Congress on December 26, 1989, and replaced the

Table 2.1 Major environmental issues in China in 1970s and 1980s.

Name of incidents	Date and place	Major reason	Serious consequences
Tangshan earthquake	At 3:42 a.m. on July 28, 1976 Tangshan	Earthquake of Magnitude 7.8	242,000 people dead, 164,000 people injured, the city ruins
Forest fire in northeast of China	May 6, 1987, Greater Khingan Range in northeast of China	Forest fires, lasted nearly a month	6.5×10^5 hm^2 natural forest burned, charred 10×10^5 hm^2, 200 people died, 226 people were injured, 50,000 people homeless, ecosystems have been seriously damaged
Shanghai HAV outbreaks	January 1988, Shanghai area	Clam seriously polluted by HAV	Shanghai citizens suffered unexplained fever, vomiting, anorexia, fatigue, jaundice, and other symptoms. Symptoms occurred suddenly, reported cases from 33 increased to 134 during one day on 18th. March 18, 1988, a total of 29,230 cases have been reported

Environmental Protection Law of the People's Republic of China (for Trial Implementation). As the basic law of environmental protection, it played a significant role (Ministry of Ecology and Environment of the People's Republic of China, 1989).

In the meanwhile, during these years, there are only a few major environmental issues, as listed in Table 2.1. Note that these are just typical representatives of China's major environmental issues, not all.

2.5 1990s—growth of environmental consciousness

2.5.1 The United States: inner consciousness

The modern environmental issues appeared when the relationship changed between human beings and the natural environment. Since the industrial revolution, human activity has fundamentally changed the resource distribution and environmental conditions at the surface of the earth. While people enjoying significantly better life qualities, more and more secondary environmental issues emerge. For instance, the artificial environment in modern cities significantly reduces the incidence of the disease, but the natural ecosystems have been destroyed. With more and more efficient and easy-to-use machines and tools, people enjoy more comfortable and convenient transportation, lots of energy consumption and air, water, energy, noise pollutions are getting closer and closer to us.

In the meantime, changing the physical environment is also getting faster and milder due to the industrial revolution. Our world, handled the environmental issues brought by the human for thousands of years, cannot catch up with this fast increase of the environmental issues, and these issues start to emerge under the public and bring serious consequence. As western culture suffered environmental issues centuries ago, nowadays, more and more people start to have an awareness to serve for the environmental issues in these countries.

In 1992, the United States had more than 10,000 nongovernmental environmental protection organizations. Members of the largest 10 organizations increased from 500,000 in 1965 to 7.2 million in 1990. For example, members of the Sierra Club were 114,000 in 1979 and reached 325,000 in 1982 (Sierra Club, 2012). The 1989 and 1990 Gallup surveys found around three-quarters (76% and 73%, respectively) of the public responding "yes" to the question "Do you consider yourself an environmentalist?" (Jones, 2016). A 1990 survey by Environment Opinion Study also found two-thirds of the public agreeing that "threats to the environment are as serious as the environmental groups say they are" and only one-quarter saying that "environmental groups are exaggerating these threats in order to get the public to pay attention to them" (Dunlap & Mertig, 1992). So, with the development of the environmental protection movement, the public's environmental consciousness was also growing.

Environmental issues have become a topic which people can hear from the media and will talk about it every day. Sociologists became providers of environmental concern information daily accessed by the public. Environmental scientists are often invited by the media to talk about science, technology, and the environment. More researchers joined and published their opinions about environmental awareness. Ruttan placed three waves of concern for natural resource implications and

Table 2.2 The three waves of concern.

Three waves time	Focus	Concerns
The late 1940s and early 1950s	The adequacy of land, water, energy, and other natural resources to sustain growth	The quantitative relationships between resource availability and economic growth
The late 1960s and early 1970s	Asbestos in our insulation, pesticides in our food, radioactive wastes in the biosphere, and the smog in the air	The capacity of the environment to assimilate the multiple forms of pollution generated by growth
The mid-1980s to 1990s	Global warming, ozone depletion, acid rain	Environmental changes that are occurring on a transnational scale

environmental change within a historical and theoretical context (Ruttan, 1993). Table 2.2 was made for a clear interpretation.

The environmental consciousness trend extended from concern in the United States to concern around the world. The US government also took the opportunity to construct programs promoting the environmental point of view to other countries and used political intervention in attempts to control the pollution and even the policy of developing countries.

2.5.2 China: reference and development

Since 1990, China has paid more attention to the environmental protection work by holding the third, fourth, fifth, and sixth national environment protection conferences. The Central government held the one-child policy and the environmental protection conferences in 1997 and 1998. From then on, population, resource, and environmental issues became routine topics of conferences in the National People's Congress (NPC). The Chinese People's Political Consultative Conference (CPPCC) put forward the transformation that turned concern mainly from pollution control to equally emphasizing attention on pollution control and ecological protection. In the 1990s, NPC and CPPCC put forward the transformation from pollution control to pollution control and ecological protection; from terminal treatment to source and entire control; from discrete point treatment to comprehensive treatment; and from density control to combine density and total control. The conference also decided to operate a clean process and circle economy, and concentrated on human resource, material resource, and finance resource on the important environmental issues area—"three rivers" (Huai River, Hai River, and Liao River), "three lakes" (Lake Tai, Lake Dian, and Lake Chao), "two areas" (sulfur dioxide control area and acid-rain control area), "one city" (Peking), and "one sea" (Bohai Sea) (Zhou, 2009).

In addition, China cooperated closely with international environmental protection. In 1990, China attended the "Basel Convention on the Control of Transboundary Movements of Hazardous Wastes and Their Disposal." In 1992, China attended the "Convention on Biological Diversity." In 1994, China attended the "United Nations Convention to Combat Desertification." In 1996, China attended the "Convention on Nuclear Safety." In 1998, China attended the "United Nations Framework Convention on Climate Change" (e.g., Kyoto Protocol), etc.

Since the "eleventh 5-year plan," the central committee promoted a requirement of accelerating the achievement of the "three historic transformations." These transformations changed policy from emphasizing economic growth but ignoring environmental protection to emphasizing both environmental protection and economic growth, with environmental protection lagging behind economic growth, to finally synchronizing environmental protection and economic growth. Moreover, these transformations changing from mainly employing administrative measures in environmental protection to the comprehensive use of legal, economic, technical, and necessary administrative measures to solve environmental problems. This requirement established a working idea: "all-around development with breakthroughs in key areas," put forward the subject that solves environmental issues from the national macrostrategic level, formulate environmental economic policies from entire process of reproduction, let overwhelmed rivers and lakes recuperated, and promote higher integration of environmental and economic policies (Zhou, 2009). During the last 20 years of the 20th century, energy consumption per unit GDP was down by 34% and greater than 1% over the last same period. China achieved its goal of triple the GDP by double energy consumption (Zeng, 2005).

Similar to the last section, a few major environmental issues during these years in China are also listed in Table 2.3, that shows the frequency of the environmental issues was slightly increased compared with the 1970s and 1980s. These are also just typical representatives of China's major environmental issues in these years.

Table 2.3 Major environmental issues in China in 1990s.

Name of incidents	Date and place	Major reason	Serious consequences
Yuanjiang dead fish	May 1991, Yuanjiang, Hunan	Phosphorus waste was discharged from three chemical factories for long term which accumulate in sediments. Heavy rains made sediment churn and released elemental phosphorus	For 40 days, large water area which across five counties suffered serious phosphorus pollution, dead fish reached 500,000 kg
Chemical warehouse explosion	August 1993, Shenzhen, Guangdong	Without approval from the environmental protection department, the warehouse stored 49 kinds of chemicals over 2800 tons. They are mostly explosive/toxic substances	Direct contact with oxidizing caused the explosion. About 15 people died, fire lasted 16 h and destroyed seven houses. Two pit reported as 20 m diameter with 9 m deep
98 Flood	Summer 1998, Yangtze River Basin	1998 summer, south China got rare heavy rain. Continuous heavy rain caused Yangtze River the largest flood since 1954	Number of million people in 29 provinces/cities/regions became refugee. Economic losses more than 1600 billion Yuan

2.6 The 21st century

2.6.1 The United States: reflections on modernity

"If the forecasting of the limits to growth theorists was full of problems, it nonetheless highlighted the truism—conveniently ignored by capitalism and its economists—that infinite expansion within a finite environment was a contradiction in terms. It thus posed a potentially catastrophic conflict between global capitalism and the global environment" (Foster, 2002).

Specifically, while the academic community of the United States reexamined the development of environmental consciousness growth into a main social concern, the recognition of challenges from environment sociology is also the symbol of environmental consciousness transformation.

The political implications made environmental policy changes during the new century. In the Clinton administration, the environmental policy obtained great support and developed both domestically and abroad. After George W. Bush became president, the environment policy experienced decreased attention. Shortly after he became president, he set aside the previous environmental laws, refusing to undertake the liability protection. President Obama differs greatly from Bush in policies concerning the environment and climate change. On January 26, 2009, President Obama signed two related presidential memorandums. In what he called "a down payment on a broader and sustained effort to reduce our dependence on foreign oil," Obama directed the Department of Transportation to establish higher fuel efficiency standards for carmakers' 2011 model year (Javers, 2009).

The global economic crisis outbreak at the end of 2008, which had its source in the United States, is inextricably linked with the environmental policy of President Obama. People believe that the reorganization of the energy industry will be a reorganization of the United States' global strategic force. It will bring the United States out of the current economic, environmental, and national security crisis. This energy revolution will let Obama successfully promote a new economic revolution primarily of a green economy. Moreover, the establishment of a clean energy structure will completely change the produce and lifestyle of the United States. The energy revolution will be based on the American Recovery and Reinvestment Act (Obama, 2009). The energy structure revolution will be the greatest economic, social, and environmental revolution in the 21st century.

2.6.2 China: mobilization of the nation

Resource-saving is an urgent task of China's development, and also is a common sustainable development demand by countries around the world. Western developed countries accumulated a lot of experience in reducing resource waste and improving the efficiency of resource use. The experience learned from the western combined with China's own national characteristics, leads China to walk in its own harmonious environment way.

Based on the thinking of previous economic growth methods and developmental paths, building a conservation and environment-friendly society and realizing sustainable development is the people's right choice to alleviate the contradiction between population, resources, and environment. At present, China has a serious environmental situation. With the industrial and urban development, the environment pressure continuously increases, and the contradiction of environmental deterioration has been developed to a critical juncture. The seriousness of resources and environmental problems in the socioeconomic system has been shown. Yue Pan, vice-minister of the Chinese Ministry of Environmental Protection gave an emphasized speech in the Fortune Global Forum, entitled "China's environmental issues are mainly caused by our distorted concept of development." In his speech, he said: "In the recent 50 years, the population of China increased from 600 million to 1.3 billion. However, the habitable land due to soil erosion reduction decreased from 6 million square kilometers to more than 3 million. One-third of the land has been acid rain polluted. Two-fifths of the major river systems have become worse than Grade V. More than 300 million rural residents are not able to get safe drinking water. More than 400 million city residents are breathing polluted air. Consequently, 15 million people get bronchial cancer or respiratory diseases" (Pan, 2005).

The minister of the Chinese Ministry of Environmental Protection Zhou believes that China's dangerous economic growth causes the increase of environmental damage. His published article wrote, in the history of civilization for thousands of years in China, the contradiction between human and nature has never been such a serious problem. The decay, deterioration, depletion, and deterioration of the ecological environment of resources becomes a bottleneck and restricts the economic and social development of China (Zhou, 2011).

A summarized table, about China's major environmental issues since 2000, is shown in Table 2.4. This table shows environmental issues outbreaks within an increased rate and confirmed the seriousness of the Chinese contemporary environmental issues. This table also shows typical representatives of China's major environmental issues, not all. For instance, from January to December 2007, pollution incidents reported by media reached 105 times.

More and more environmental problems appear in China, however, improving the quality of the environment is a long-term and arduous process. To address this issue, the Chinese government put forward the concept of building an environment-friendly society (Hu, 2005). In order to accelerate building the environment-friendly society, the Ninth National People's Congress Standing Committee 30th meeting accessed "Law of

Table 2.4 Major environmental issues in China since 2000.

Name of incidents	Date and place	Major reason	Serious consequences
"9.29" Danjiang sodium cyanide pollution	September 2000, Danfeng County, Shanxi	Unqualified vehicles transported 10.33 tons of sodium cyanide, overturned in the halfway	Caused Danjiang serious sodium cyanide pollution. Direct economic losses reported 11.88 million Yuan
SARS epidemic	From November 2002 to August 2003, China and worldwide	Viral respiratory illness caused by a coronavirus	November 16, 2002, the first cases appeared in China. The disease has a strong infectious virus which can severely damage the human body and leads to death. China reported a total of 5,327 cases, 349 died
Chongqing Kaixian blowout accident	22:15 p.m. December 23, 2003, Gaoqiao Town, Kaixian County of Chongqing	Oil drilling company Chuandongbeiqi gas field 16H well got suddenly blowouts during drilling. Hydrogen sulfide gas ejected up to 30 m high	Toxic gas rapidly spread in the air and caused a major disaster. 243 deaths and More than 4,000 injured. More than 60,000 been evacuated
Tuojiang water Pollution	March 2004, Tuojiang, Sichuan	Fertilizer plant discharged high concentrations wastewater into Tuojiang	Drinking water interruption for 26 days. Impacted nearly a million people and a large number of enterprises. The direct economic losses about $300 million Yuan. Tuojiang's ecological system cannot be recovered for less than 5 years.
Songhua Jiang water pollution	November 2005, Jilin City, Jilin Province	Benzene plant explored, hundreds of tons of nitrobenzene and other toxic chemicals polluted the Songhuajiang water	About 8 died and 70 injured because of the exploration. Water pollution into 135 km. Drinking water for Harbin city has been cut off for 4 days, which impacted millions of people

(Continued)

Table 2.4 (Continued)

Name of incidents	Date and place	Major reason	Serious consequences
Yueyang Arsenic Contamination	September 2006, Yueyang, Hunan	Drinking water protection area got polluted by arsenic	Arsenic has been found 10 times higher than the standards. It impacted 80,000 residents
Taihu Lake Cyanobacteria Break-out	May 2007, Taihu, Wuxi	A large area of Taihu lake suddenly occurred cyanobacteria Break-out under high-temperature conditions	Drinking water quickly been contaminated. Wuxi suffered a water shortage
Sanlu scandal milk	September 2008. A number of provinces simultaneously broke	Sanlu Group Co., Ltd. Production, Sanlu infant formula powder milk contaminated with melamine	"Kidney stones' baby" appeared throughout the country within 2 months. Gansu 59 cases, 1 death; Jiangsu 20 cases; Hubei 3 cases; Shaanxi 2 cases; Anhui 3 cases. Shandong, Jiangxi, Ningxia and other provinces also found similar cases
Yancheng water crisis	February 2009, Yancheng, Jiangsu	Morning on February 20, the water supply plant contaminated by phenol.	Two water plants had been shut down. Yancheng suffered a water shortage Bottled water price has been raised twice in 10 min. It impacted 200,000 residents
Xingang pipeline bombings	July 16, 2010, Xingang, Dalian	Dalian Xingang oil pipeline exploded. The preliminary estimate of discharged oil was 1500 tons	Chinese ocean surveillance ship monitoring at 13:30 on the 19th and reported 430 km^2 pollution area, with heavily pollution about 12 and 52 km^2 general pollution area

the People's Republic of China on the Environmental Impact Assessment" on October 28, 2002. The State Council issued a decision that implements the scientific development view and enhanced environmental protection on December 3, 2005. "Creating a resource-saving and environment-friendly society" development strategy, in essence, is a value that concerns people's livelihood and social development.

In 2007, the Organisation for Economic Co-operation and Development (OECD) published OECD Environmental Performance Reviews: China. In this review, it is believed: that the "efforts (regulatory, economic instruments, and policy) have neither been sufficient to keep pace with the environmental pressures and challenges generated by the very rapid growth of China's developing economy nor to capture the potential economic benefits to be obtained from improved pollution abatement and nature protection. Overall, environmental efforts have lacked effectiveness and efficiency, largely as a result of an implementation gap. The weaknesses in the present system are demonstrated by the failure to achieve some of the key objectives of the 10th FYP, and the severity of environmental problems in many parts of China" (Organisation for Economic Co-operation and Development, 2007). In addition, the Chinese Public Environmental Protection Index by the China Environment Culture Promotion Association (CECPA) is considered as the Chinese public environmental protection awareness and behavior's "barometer" and it is also the first environmental protection annual index. In 2008, the environmental pollution problem ranked third in "public most concerned social hot topics." It is the third time the environmental pollution problem entered the top three rankings. Secretary-General of CECPA Shaomin Zhang said: "Environmental pollution problems closely follow the 'price issue' and 'food security.' The public concern rate is 37.7%. Environmental pollution and various environmental events still exist as threats to the socioeconomic sustainable development and personal life of the public. Some of them are very serious. The environmental problems have risen to the height and level which impact of social harmony and stability." Facing the serious environmental situation, the Chinese government did not give up efforts but instead used the 2008 Olympic Games as an opportunity to increase environmental protection efforts and enhance environmental consciousness.

"Green Olympics" was an important vow taken by Beijing to the International Olympic Committee. The so-called "Green Olympics" was goals of comprehensive movement to enhance the city's sustainable development capacity, provide a clean and beautiful environment for the

Table 2.5 The concept historic evolution of the relationship between human and nature (Cai, 1989).

	The hunter-gatherers civilization	The agricultural civilization	The industrial civilization	The post-industrial society
Relationship with nature	Worship of nature	Transformation of nature	Conquest of nature	Seeking human nature coordinate
Productivity level	Low productivity, increase slowly	Continued to rise	Rapid increase due to the technical rapid development	Continuously increase at an astonishing rate
Human activities	In the hunter-gatherers civilization, feed on animals and plants	Into the agricultural civilization, development and utilization of land, water, climate and other resources	Entered the industrial civilization, attempt to dominate nature, sacrificing nature, accumulate the wealth	Re-examine the economic behavior, concern the environment and development issues
Human nature relationship	Fear and dependence	Dependence weakened, confrontational enhanced	Uncoordinated, people and nature conflicts intensified	Seek the coordination of resources, environment and development
Population	Limited	Start rise	Fast rise	Rapid rise continuously
Environmental issues	Nothing serious	Geographical environment worsening	Environmental pollution, ecological damage, endanger the survival of mankind	Flood hazards, humans began to realize that environmental issues, and gradually solve environmental problems

Olympic Games 2008, organize the Olympic Games with sustainable development perception that protect the environment, conserve resources and protect ecological balance, minimize the Olympic Games' negative impacts to the environment, play an exemplary role for environmental protection work all over the country, carry out environmental education activities extensively to enhance public environmental consciousness, emphasis on harmony purposes between man and nature, and comprehensive enhancement of public ecological civilization literacy (The official website of the Beijing 2008 Olympic Games, 2008).

In fact, the Olympic Game was like a "catalyst." With the policy, funding, and public support, environmental protection concepts had been fully released. In addition to the policy change, individuals all changed their behaviors to protect the environment. At the same time, the Olympics provided an opportunity to apply the laboratory phase environmental protection technology and environmental protection products. For example, in the process of building the Olympic venues, many new technologies were utilized including stormwater utilization, solar photovoltaic, the light tube system engineering, ecologic water system, recycled water utilization, high efficient energy-saving lamps, water-saving toilets, new energy utilization, and solid waste collection and treatment. This collection of technologies of environmental protection used in the Olympic Center's facilities can be seen as "Fair" for environmental protection design. There is no doubt that Chinese environmental consciousness and environmental management were significantly increased by the Olympic Games. Environmental policies and systems were becoming better and approaching perfection day by day. Although it is difficult to pass on environmentalism as a heritage, the Chinese will try their best to do so. Director of Beijing Municipal Environmental Protection Bureau Shaozhong Du said that "the reduced pollution measures will be certain as the fixed form and even more stringent after the Olympic Games. Beijing, and other cities will play an exemplary role and have a durable influence in the coming days" (Zha, 2008).

At the end of this chapter, we review the environmental issues on the timeline. The evolution of the relationship between humans and the natural environment is demonstrated more clearly. In 1989, Yunlong Cai of China systematically summarized the concept historic evolution of the relationship between humans and nature in the hunter-gatherers' civilization, agricultural civilization, industrial civilization, and postindustrial society (Cai, 1989), as shown in Table 2.5.

CHAPTER THREE

From movements to measurement

3.1 The movement trend—The United States and China

Through a diachronic comparison of the environmental consciousness in the United States and in China in the last chapter, we can see that awareness of environmental issues rose rapidly in the United States from the 1960s to the early 1970s. That was around the time that the first wave of environmental consciousness had been formed in the society with the first wave of industrial civilization. During that time, the establishment of heavy industry with coal, metallurgy, and chemical in western society, economics rose fast in those countries; however, environmental problems with pollutions and nature resources deterioration became more and more serious. As a short summary of the previous chapters, Meuse Valley fog killed 60 people in Belgium due to a combination of industrial air pollution and climatic conditions in 1930 (Nemery, Hoet, & Nemmar, 2001); Donora smog killed 20 people and sickened 7000 more in Donora, Pennsylvania, the United States in 1948; then in 1952, "Great smog" or "Big smoke" affected London, 4000 people had died prematurely and 100,000 more were made ill because of the smog's effects on the human respiratory tract (Seinfeld, 2004); in 1956, Minamata disease was first discovered in Minamata city in Kumamoto prefecture, Japan and 1784 had died (Environmental Health Department, n.d.).

Those continuing serveries environmental issues triggered the consciousness changing in human society. From the beginning of the 1960s, the modern concept of environmental awareness is developed based on this background. In 1962, the American marine biologist Rachel Carson published her book, "Silent Spring," the book then is recognized as the opening of the worldwide "ecological era." After "Silent Spring," more and more people took serious considerations in their living environment. It made a big movement in western society, and impacts the change in

Environmental Consciousness in China.
DOI: https://doi.org/10.1016/B978-0-08-100388-6.00003-9

policy, thoughts, lifestyle, and even on the beliefs. All aspects of life set off a wave of "green environment," and formed the concept of the word "environmental consciousness/awareness."

In that decade "primary attention was given to air and water pollution and the importance of protecting areas of natural beauty and recreational value" (Dunlap & Marshall, 2007). During the period from 1973 to 1974, energy crises made "The Limits to Growth" a guiding principle. Chemical waste made sociological attention turned to local environmental hazards and introduced the different functions of environmental protection to public awareness. However, rainforest destruction and the decline of biological diversity turned the public attention back to the global level. Furthermore, global warming increasingly made environmental concerns more intensive. Table 3.1 summarized the development of environmental consciousness in the United States.

In Hannigan's opinion, the role of two major societal institutions—science and the mass media—plays an important role in constructing environmental risks, knowledge, crises, and solutions (Hannigan, 1995). This point is verified by the environmental consciousness development process of the United States.

Professor Dunlap supported Hannigan's view that Environmental Concern came from the deterioration of the environment. But Silent Spring brought the reflection of environmental issues to people of the United States. Mass media promoted environmental consciousness causing a peak in the early 1970s. Then came the recession (explanation to follow); however, mass environmental consciousness had already been established. Based on this, mainstream environmental concerns impelled the government to focus on environmental protection when the policy became slack.

In China, the first reason for environmental concern is also the deterioration of the environment—"1950s wash rice and vegetable, 1960s wash

Table 3.1 The development of environmental consciousness of the United States.

1960s	Developed rapidly
The first Earth Day on 1970s	Peaked
Later 1970s	Declined but better than mid-sixties
1980s	Backlash
1990s	Boost and higher than early seventies
2000s	Structural reflection after holding new consciousness

clothes and irrigation, 1970s water turned bad, 1980s fish and shrimp vanished, and 1990s harmed the human"—this folk song prevails over the Huaihe River area which reflects water pollution history vividly. Based on those environmental issues, the government established laws and regulations for environmental protection. However, the consciousness of environmental protection has not struck root in the heart of people. When situation worsened, local people took action to protect the environment and formed nongovernmental organizations (NGOs) spontaneously (such as Friends of Nature). Although the influence is far below the Rome Club which impelled the environmental protection progress of the United Nations, the efforts did contribute to an awakening of environmental consciousness. The author believes that the Chinese environmental consciousness influenced by the government environmental top–down protection system on the one hand. On the other hand, Western science, mass media, and the impact of the United Nations also played an important role in raising environmental consciousness in China. In addition, the spontaneous discussions and actions by some NGOs have also promoted the awareness of the Chinese environmental consciousness.

By comparing Tables 3.1 and 3.2, we can see that the environmental consciousness started late in China. The awakening of the environmental consciousness was followed by the United States and other developed countries, but the overall tendency of environmental consciousness is increasing and close to maturity.

3.2 The measurement tool

For a deeper understanding of the change of the environmental consciousness in China, including current status, the path of development, and the future tendency, the research of environmental consciousness or

Table 3.2 The development of environmental consciousness in China.

1960s	No environmental consciousness
The first Earth Day on 1970s	Environmental consciousness awareness
Later 1970s	Development of environmental consciousness
1980s	Keep pace with international situation
1990s	Became important
2000s	Flourishing

awareness has been performed. In the previous chapters, simple questions such as "Do you consider yourself an environmentalist?" or "environmental groups are exaggerating these threats in order to get the public to pay attention to them" are used to measure the environmental consciousness. However these simple questions cannot fully reflect the environmental consciousness in all aspects.

In the 1960s, western researchers began the systematic study of the relationship between human behavior and the natural environment. Since then, this new trend also fashions in the sociology arena. Sociology researchers hope to learn more about the relationship between humans and the environment.

In the early works of Catton and Dunlap, they brought out the narrow anthropocentrism of classical sociology (Catton & Dunlap, 1978). Their published research proved to be a watershed moment in environmental sociology, refocusing research in the field and marking a new epoch. They thought that the traditional type of sociological research, modeled after Durkheim, neglected the nature of environmental influence on human society. They pointed out that the foundation of the traditional western worldview is that human beings are the masters of the world, using the resources of the world to deal with current and developing problems. After Human Exemptionalism Paradigm was challenged by the New Ecological Paradigm, the transition was made from a Sociology of Environment to an Environmental Sociology.

Many more researchers followed their path and contrasted traditional sociological thought and emerging environmental sociology. They tried to explore a sustainable approach to social development (Gardner & Stern, 2002). Sociologists wanted to know, why people treat the environment like the way they are doing now. To achieve this goal, many scholars made great efforts. By the end of 2000, more than 1000 articles of ecological consciousness have been published (Dunlap & Jones, 2002).

With the growing number of empirical studies on environmental consciousness, scholars figure out that development of an effective tool to measure environmental consciousness is crucial in this study. Otherwise, reliability and validity are difficult to identify, and research findings are hard to compare. If people want to heap up the knowledge of environmental consciousness, a valid and reliable scale for measuring eco-consciousness is essential.

Despite many scales, only three scales are widely used (Fransson & Gärling, 1999). They are New Ecological Paradigm Scale (NEP) (Dunlap

& van Liere, 1978), Ecological Attitudes and Knowledge Scale (Maloney & Ward, 1973), and Environmental Concern Scale (Weigel & Weigel, 1978). Ecological Attitudes and Knowledge Scale is derived from psychology research and is one of the most widely used scales to measure the environmental consciousness in American academic. The origin version of the scale contains about 130 questions. Table 3.3 listed some typical questions. These questions are further divided into four groups including verbal commitment, actual commitment, affect, and knowledge. This scale is characterized by measuring multiple environmental topics from multiple levels, with comprehensive aspects; however, it is too complicated and it is believed that the scale is lacking deep examination on the face and/or content validity (Dunlap & Jones, 2002).

As an academic groundbreaking achievement, NEP scale is considered as a widely used method of measurement of ecological consciousness (Cordano, Welcomer, & Scherer, 2003; Stern, Dietz, & Guagnano, 1995; Zelezny, Chua, & Aldrich, 2000). Furthermore, Stern et al. even repute that the NEP scale is the most widely used scale to the 1990s (Stern et al., 1995). The literature review also found that there are still a lot of people who use this scale in the 2000s (Hunter & Rinner, 2004). NEP scale is not limited to academics, and not confined to the United States. In fact,

Table 3.3 Typical questions in Ecological Attitudes and Knowledge Scale (Maloney & Ward, 1973).

1. I'd be willing to ride a bicycle or take the bus to work in order to reduce air pollution.
2. I would probably never join a group or club which is concerned solely with ecological problems.
3–36. 34 more questions related to the verbal commitment subscale are omitted.
37. I have switched products for ecological reasons.
38. I save some waste materials for recycling.
39–72. 34 more questions related to the actual commitment subscale are omitted.
73. I become incensed when I think about the harm being done to plant and animal life by pollution.
74. I get depressed on smoggy days.
75–106. 32 more questions related to the affect subscale are omitted.
107. Which of the following materials usually takes longest to decompose? (A) Tin, (B) Iron, (C) Aluminum, (D) Copper, (E) Steel.
108. Mercury has been found in unacceptable levels in (A) Fruit, (B) Vegetables, (C) Seafood, (D) Beef, (E) Soft drinks.
109–130. 22 more questions related to the knowledge subscale are omitted.

many countries, academics, civil society, and government agencies are using the NEP scale.

Dunlap et al. set 15 items for the NEP scale as shown in Table 3.4 (Dunlap, Liere, Mertig, & Jones, 2000). Three items were designed to tap each of the five hypothesized facets of an ecological worldview: the reality of limits to growth (1, 6, 11), antianthropocentrism (2, 7, 12), the fragility of nature's balance (3, 8, 13), rejection of exemptionalism (4, 9, 14), and the possibility of an ecocrisis (5, 10, 15). The eight odd-numbered items were worded so that agreement indicates a proecological view, and the seven even-numbered ones so that disagreement indicates a proecological worldview.

In China, Dayong Hong first discussed the NEP scale in Dunlap and Cottan's paper in 1999 (Hong, 1999). Since then, people's attention to environmental sociology has gradually increased, and the discussion on the paradigm of environmental sociology has also increased (Li, 2000; Lu, 2004; Xu, 2006). At the same time, with the deterioration of domestic environmental pollution, discussions on studies of the specific relationship between environment and society are increased, the study of environmental consciousness has also attracted more and more interest, and the NEP

Table 3.4 List of question in New Ecological Paradigm (Dunlap et al., 2000).

1. We are approaching the limit of the number of people the Earth can support.
2. Humans have the right to modify the natural environment to suit their needs.
3. When humans interfere with nature it often produces disastrous consequences.
4. Human ingenuity will insure that we do not make the Earth unlivable.
5. Humans are seriously abusing the environment.
6. The Earth has plenty of natural resources if we just learn how to develop them.
7. Plants and animals have as much right as humans to exist.
8. The balance of nature is strong enough to cope with the impacts of modern industrial nations.
9. Despite our special abilities, humans are still subject to the laws of nature.
10. The so-called "ecological crisis" facing humankind has been greatly exaggerated.
11. The Earth is like a spaceship with very limited room and resources.
12. Humans were meant to rule over the rest of nature.
13. The balance of nature is very delicate and easily upset.
14. Humans will eventually learn enough about how nature works to be able to control it.
15. If things continue on their present course, we will soon experience a major ecological catastrophe.

scale was used in China. In 2003, NEP scale was used in the Chinese General Social Survey (CGSS) and obtained relevant data. In 2007, Dayong Hong and Chenyang Xiao used the CGSS 2003 data to test the reliability and validity of the NEP scale, and believed that it can be used as an important tool for measuring public environmental concerns in China with a minor modification (Hong & Xiao, 2007). Thereafter, the NEP scale has gradually been applied to the study of environmental consciousness in China by more and more researchers in China and has received extensive attention.

In 2007, the NEP scale was once again reflected as an important tool for measuring the differences in environmental consciousness between the genders in China by Hong and Xiao (2007). In 2008, Guang Yang incorporated the measurement of the NEP scale into the scale of teacher's belief in China (Yang, 2008). In 2010, the NEP scale was included in China's public environmental literacy assessment indicators by the research project team (China's Public Environmental Literacy Assessment Indicators Research Project Team, 2010). In 2011, Chuntian Lu and Dayong Hong performed a detailed analysis of the possibility of in-depth research on the NEP scale in China (Lu & Hong, 2011). These show the broad use of the NEP scale in research in China.

3.3 Measure the environmental consciousness

In this book, we follow the NEP scale as a quantitative measurement of the environmental consciousness in China and use a translated version of NEP scale in our questionnaire. The translation quality may have a large effect on the result of the questionnaire due to the potentially ambiguous or obstacles in understanding. To solve this issue, we used the Chinese translation version by Dayong Hong in the research (Hong, 2005).

In 2009–11, the author conducted the survey in three field areas, that is, Nanjing (the capital city of Jiangsu province), Nanjing city suburbs, and rural areas of northern Jiangsu, Anhui, Henan, and Shandong provinces. In the urban area of Nanjing, the stratified sampling method was mainly used for the survey. The places to conduct the survey are widely distributed in the urban area of Nanjing as shown in Fig. 3.1. Three levels of the living circles are included, that is, high-end living circle (Longjiang Community), the mid-range living circle (Jingming Garden), and the

Figure 3.1 Map view of the places to conduct the questionnaire survey of urban residents.

low-end living circle (Yuhua New Village). Three levels of densely populated areas were also included, that is, high-end consumption center (Xinjiekou), mid-range consumption area (Confucius Temple), and low-end consumption area (Central Gate Station). In order to balance the age ratio, a supplementary survey was conducted in Jiangnan Youth City and Xianlin University City where young people lived in a concentrated manner. The stratified sampling method mitigates the defects in the occasional sampling method and ensures the validity of the sample data.

For rural areas, the authors mainly conducted the questionnaire survey, interviews, and participatory observations in northern Jiangsu. In addition, questionnaire surveys were also conducted in rural areas in Anhui, Henan, and Shandong Provinces. The specific survey takes the form of both personal interviews and interviews by questionnaire investigators. The proportions of questionnaires issued by rural surveys are 45% in the northern part of Jiangsu, 25% in Anhui, 12% in Shandong, 12% in Henan, and 6%

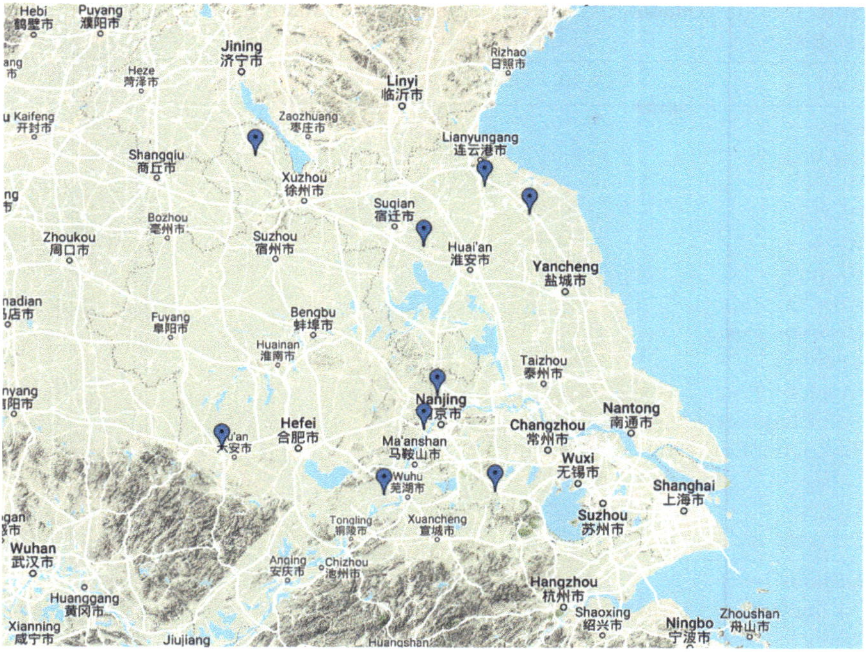

Figure 3.2 Map view of the major places to conduct the questionnaire survey of rural residents.

in other provinces. Fig. 3.2 shows the major places where the questionnaire survey in rural areas were conducted.

Questionnaires, interviews with participatory observations were conducted during the research. After completion of the investigation, we recode score according to the dimensions of the term, that is, 1, 3, 5, 7, 9, 11, 13, 15 terms are positive dimension, respondents answered strongly agree, agree, uncertain, disagree, strongly disagree were recoded as 5, 4, 3, 2, 1; for negative dimension, including 2, 4, 6, 8, 10, 12, 14 terms, respondents answered strongly agree, agree, uncertain, disagree, strong disagree were recoded as 1, 2, 3, 4, 5. Thus, the scale score range of the NEP scale is from 15 to 75 points.

A total of 1067 questionnaires were retrieved from this survey, and 1008 of the questionnaires were valid. From the perspective of important demographic variables, the distribution of factors such as gender, age, education level, religious belief, and survey location is reasonable, as shown in Table 3.5.

Table 3.5 Distribution of basic factors of the valid questionnaires.

Gender			
Male	54.5%	Female	45.5%
Age			
Less than 20			8.2%
21–30			42.1%
31–40			21.4%
41–50			16.3%
51–60			6.1%
More than 60			6.0%
Education level			
Illiteracy			3.8%
Primary school			6.4%
Junior high school			16.9%
High school or secondary school			25.1%
College			19.1%
Bachelor			23.8%
Graduate and above			4.8%
Religious belief			
Do not have	88.5%	Have	11.5%
Survey location			
City	77.1%	Rural area	22.9%

3.4 Environmental consciousness in China

The histogram of NEP scores in the survey is shown in Fig. 3.3. For all the valid 1008 questionnaires, the lowest score is 33 and the highest score is 75. The average of the NEP score is 55.9 with a standard deviation of 8.06. The NEP scale of the survey takes place in the city and rural areas are first compared. The average NEP score for the survey takes place in the city is 57.55 with the standard deviation at 7.41 and in the rural areas is 50.28 with a standard deviation at 7.66. Fig. 3.3 shows the histogram of the data comparing the survey take place in the city and rural areas. It is straightforward to see that the NEP of the survey take place in the city is significantly larger than in rural areas.

This result is consistent with the previous researches that urban–rural dichotomy is dominant in sociology surveys in China and urban residents have higher ecological consciousness levels than rural residents. For example, a nationwide survey of public ecological consciousness organized by

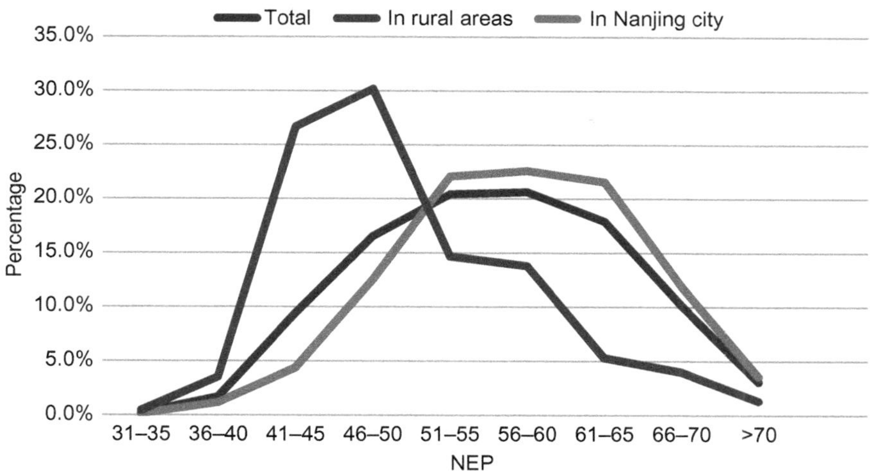

Figure 3.3 Histogram of NEP scores of the survey conducted in city and rural areas.

the Ministry of Environmental Protection and the Ministry of Education in 1998 found that the environmental consciousness level of rural people was much higher than urban people. Environment knowledge test shows that the national average score was 2.8, urban residents got 4.5 on average while rural residents only got 2.4 on average (Xu, 1997). Another continued data from the Chinese public Environment and Inhabitant livelihood Index (2005–08) showed that people from different areas have an extremely significant difference in ecological consciousness. The ecological consciousness of urban residents is 68, which was significantly higher (7.5 points higher) than rural residents in China (China Environmental Culture Promotion Association, 2005–2009). Other articles also confirmed that the average difference in the scores between urban and rural residents is very large (Gao & Meng, 2011; Yan et al., 2010).

Our result is then compared with the result in Dayong Hong's article (Hong, 2005). In Hong's article, the NEP scale value was based on the measurement of the public environmental consciousness of the urban residents of the whole country in 2003 as part of the 2003 CGSS. The results showed that the average of the environmental awareness of the 4994 respondents was 61.24 points with a standard deviation at 12.07. Here, a different scale with the minimum score at 0 and the maximum score of 100 is used. From all the respondents, the lowest score is 13.33 and the highest score is 100, with 19.8% of the respondents for less than 50 points,

33.1% for 50–59 points, 25.7% for 60–69 points, 14.6% for 70–79 points, and 6.8% for 80 points or higher.

In order to facilitate the comparison, there are three major differences between the two surveys. First, the minimum and maximum NEP score in Hong's paper (this study) is 0 and 100 (15 and 75), respectively. To make the data comparable, we convert the results of this study into the same scale with Hong's paper only in this section. After scaling the data, the lowest score and the highest score changes to 30 and 100, respectively. The average of the NEP scale value changes to 68.17 with a standard deviation of 13.43.

Next, the geographic regions of the surveys are different. Hong's research covered as CGSS only including the urban residents of most cities in China, while our research focuses on Nanjing city and surrounding areas. Fortunately, Hong's research studied the NEP by the level of the cities. He grouped the cities into three levels, direct-controlled municipality cities, capital cities of provinces, and other cities. By considering Nanjing is the capital city of Jiangsu province, it is an extra-large city based on China city size division standard (State Council of the People's Republic of China, 2014), and the population of Nanjing is around eight million, ranked around 30 in all cities in China. Fig. 3.4 shows the city size of Nanjing based on the size division standard of cities in China.. The NEP scale of the survey takes place in Nanjing, a typical capital city of provinces in China may thus represent the NEP of the urban residents in capital cities of provinces in China. As a result, the NEP scale result of the urban residents of the capital cities of provinces in Hong's paper is compared with the respondents in Nanjing.

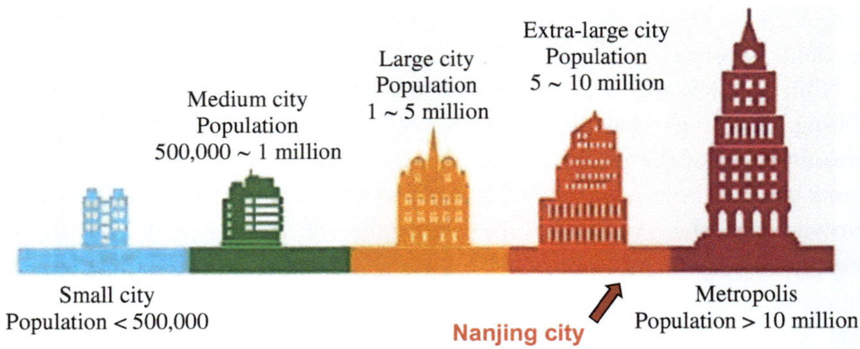

Figure 3.4 City size of Nanjing based on the size division standard of cities in China.

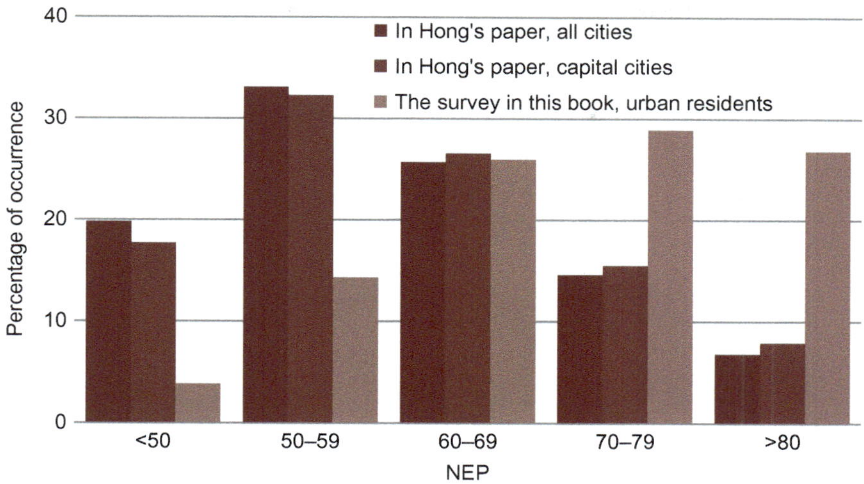

Figure 3.5 Comparison of the NEP score between this research and Hong's paper (Hong, 2005).

Lastly, the survey of this research is taken more than seven years later than the survey studied in Hong's paper. Therefore, after we made the first two adjustments, the difference of the NEP between the two surveys should represent the NEP changes of the urban residents in China over these seven more years. In Fig. 3.5, the NEP scores between Hong's paper (Hong, 2005) and our survey conducted in 2010 are compared. During these over seven years, the average of the NEP of urban residents was increased from a value less than 66.3 to 70.92 and highly scored fraction grown very fast, as shown in Fig. 3.5. Here, the average NEP of the residents of the capital cities of provinces in Hong's paper is not given and the highest possible mean NEP score is estimated by assuming the highest NEP score in the corresponding score range for each respondent. This finding is consistent with previous research conclusions that the level of environmental consciousness has increased rapidly in the urban area in China (China Environmental Culture Promotion Association, 2005–2009).

CHAPTER FOUR

From a longitudinal study to cross-sectional study

4.1 Research design and hypothesize

In the last chapter, two important conclusions were found from the result of our survey: the environmental consciousness of urban residents has increased year by year, and the weak environmental consciousness of rural residents compared with the urban residents. These two main conclusions are consistent with the previous researches; however, these two conclusions seem as a paradox if the rapid urbanization scenario in China is considered. Before we see the paradox in detail, let us see the rapid urbanization in China first.

The urbanization changed Chinese society tremendously. During the economic growth of China in the past 30 years, the structure of the society in China changed deeply. It went through a rapid industrialization and urbanization stage. From the 1980s, China's rural–urban immigration flow became bigger and bigger. As showing in Fig. 4.1, in 1990, there were only 21.25 million migrants from rural areas to cities. Within 5 years, the number bloated to 70.73 million. Census 2000 states that there were 102.29 million migrants all over the country. As of the year 2010, the population of migrants achieved 261.39 million (Population censuses). This number is more than 10 times larger than 20 years ago, that is, 1990. Compared with the migrants, there are 691 million urban residents in China in 2011. The migrants became more than a third of the total urban residents, which largely changed the population structure in cities.

The rapid urbanization brings a huge amount of people resettled from rural areas to urban cities. In the meantime, China was domestic of agriculture in the rural area and these residents have low environment consciousness in the past years. When these residents moved to cities, these rural residents with low environmental consciousness may lower the environmental consciousness of the urban residents after the resettling, which

Environmental Consciousness in China.
DOI: https://doi.org/10.1016/B978-0-08-100388-6.00004-0

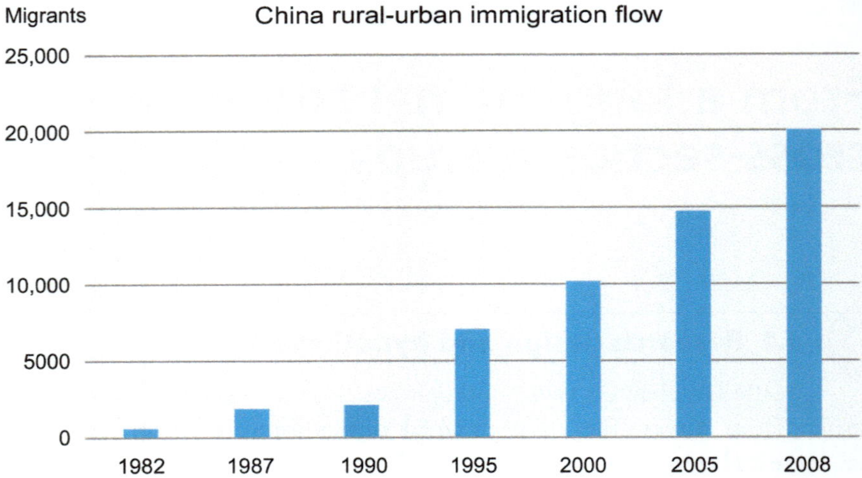

Figure 4.1 The population of rural–urban migrants in China in unit of ten thousand (Population censuses).

includes the original urban residents and the migrants. However, as discussed in the last chapter, the environmental consciousness of the new urban residents is increasing year by year, which becomes a paradox as discussed at the beginning of this chapter.

To solve this paradox, it is needed to consider the urbanization that is not only the rural people moving to cities and not only just the change in the physical aspect, but also more importantly, urbanization is about the change in the urban social structure, the change of social values and cultural norms, where the environmental consciousness is one of them. The environmental consciousness of the urban residents after the resettling may not be simply treated by combining the migrants and original urban residents. To better understand the environmental consciousness changes after the resettlement, the environmental consciousness of both urban and rural residents are studied. In previous research, these two groups of residents are rarely included in one research, thus hardly finding explanations on the change of the environmental consciousness. In addition, to understanding "the change," a longitudinal study is usually used in clinical researches.

This study layout an ambitious design based on China's rapid urbanization scenario. Based on this, the author has germinated the environmental consciousness from the fixed location and time to the broader dimension of time and locations. In the methodology, it mainly adopts the continuous spectral analogy research method. It takes cross-sectional studies instead of longitudinal studies, which take advantage of recognizing the whole picture, while not missing the timelines comparison, as the special large amount of migrants in China offers the possibility. Using the above research method, this research considered the environmental consciousness of rural residents on the early timeline, while the environmental consciousness of urban residents after the resettlement on the later timeline since previous research shows that the ecological awareness in cities increased year by year, but it increased slowly in rural area is considered (China Environmental Culture Promotion Association, 2005–2009). In addition, the urban residents after the resettlement are split into original urban residents and migrants, and three groups of residents are investigated: rural residents, urban residents, and the migration group. By doing this, this ambitious design helps to understand the change of the environmental consciousness in China and addresses the gap of previous research to understand why the environmental consciousness is different between urban and rural residents by considering the following few questions: Are the environmental consciousness of urban residents and immigrants in cities at the same time and location the same? Similarly, do rural residents and migrants have differences in their environmental consciousness? Finally, by studying the change of the environmental consciousness of the migrants, effective ways to improve the environmental consciousness may be found.

4.2 Three groups of residents

To answer the above questions, a typical city, which the origin rural area of the immigrants can be easily determined, would be an ideal sample for our research. Yangtze River Delta is an important industrial and agricultural area, and one of the most vital economically dynamic region in China. This area holds the highest population density, highest level of urbanization, and fastest-growing economic as well. As one of the core cities in the Yangtze River Delta and the provincial capital of Jiangsu, Nanjing is a good mirror to show the urbanization status of China. The migration ratio was 19.4% and the total population now was 8,004,680 in 2011 (National Bureau of Statistics of China, 2010). On the other hand, more than 70% of Nanjing migrations came from North Jiangsu Province and Anhui Provence (Cao, 2011; Zhu & Chen, 2004). The comparison between the migrants and the rural area where the migrants came from can thus be easily performed. With this, Nanjing was selected to be the typical city.

As introduced in the last chapter, the data present in this book are all from a survey investigated in 2010–2011. The respondents of the rural areas are spectrum based on the provinces where the Nanjing migrants from according to Zhu and Chen's book and a report by the leading group office of migrant workers in Jiangsu Province (The Leading Group Office of Migrant Workers in Jiangsu Province, 2007; Zhu & Chen, 2004). Zhu and Chen point out that in 2005, the migrants of Nanjing had reached 1 million, which is 24% of the total urban population (Zhu & Chen, 2004). About 42.38% of the migrants are intraprovincial migrants mainly coming from the northern part of Jiangsu. The interprovincial migrants mainly come from Anhui, Sichuan, Shandong, Zhejiang, Henan, Hubei, and Guizhou Provinces. About 60.5% of the interprovincial migrants are from Anhui Province. In another report, the leading group office of migrant workers in Jiangsu Province also mentioned that the migrants in Nanjing are mainly from Jiangsu, Anhui, Henan, Sichuan, Shandong, Guizhou, Hubei, Jiangxi, and Zhejiang Provinces (The Leading Group Office of Migrant Workers in Jiangsu Province, 2007). The migrants from the north of Jiangsu Province accounts for more than 40% of the migrants in Nanjing. In addition, migrants from Anhui, Henan, Sichuan, and Shandong account for 23.2%, 7.7%, 4.6%, and 4.5%, respectively. The migrant fraction from all other provinces is less than 13%. Based on these fractions, the location of the surveys for rural residents are thus selected as 45% in northern part Jiangsu, 25% in Anhui, 12% in Shandong, 12% in Henan, and 6% in other provinces. The major locations of the rural surveys are shown in Fig. 4.2 in the last chapter.

To filter out the migrants in the survey, we designed two questions. In the questionnaires for urban residents, each respondent was asked his or her age and living time in the city. If the two values are equal, he or she will be classified as an urban local person. Otherwise, he or she will be classified as an immigrant. The second question was asking immigrants where they came from. If "rural" is chosen as where they came from, they are identified as immigrants. Otherwise, the questionnaires are excluded. Questions with similar logic were also used in questionnaires for the rural residents. The responders from rural areas were filtered to

Figure 4.2 City migrant workers. https://commons.wikimedia.org/wiki/File:Migrant_worker_of_china.jpg

Table 4.1 Summarize of the filter logic of three groups of residents.

	Respondents in the city	Respondents in rural areas
Never migrated	Urban residents	Rural residents
Migrated from rural areas to cities	Migrants group	Migrants group
All others	Excluded	Excluded

identify the rural residents who have emigration experience. The logic of the filtering is summarized in Table 4.1.

We combined the immigrants in the urban respondents and the emigrants in the rural respondents to the "Migrants Group." The native Nanjing citizens in the urban respondents become the "Urban Residents Group," and native rural inhabitants of the rural respondents from all the provinces together as the "Rural Residents Group."

Among the valid 1008 questionnaires discussed in the last chapter, 259 samples are characterized as the Urban Resident Group and 176 samples are characterized as the Rural Residents Group. Migrant Group is our important survey group, so we did many more samples than other groups, in a total of 381 samples. The 192 samples are migrants, however, are not from rural to urban, we thus excluded them. As a result, 816 samples are within our interest groups and were analyzed. Characteristics of samples were basically consistent with general statistic results except for Rural Residents Group since the samples are from several different villages and its representativeness is unknown. Therefore, the Urban and Migrants Groups should have good representativeness. Samples of Urban Residents Group are consistent with population characteristics of Nanjing City and samples of Migrants Group showed its own traits, such as the younger age and the higher male ratio, which agrees with previous studies of migrants, as shown in Table 4.2 (Zhu & Chen, 2004).

4.3 Different consciousness level

After sorting and coding, survey data were imported into SPSS 18.0 for correction and statistical analysis. Frequency analysis was used for checking the representativeness of the data. Besides, data descriptions also give us an overview of each dimension, as shown in Table 4.2. Then,

Table 4.2 Distribution of basic factors of three groups of residents.

	Urban Residents	Rural Residents	Migrants Group
Male	47.1%	51.8%	58.3%
Female	52.9%	48.2%	41.7%
Mode of the age	27	20	25
Mode of the education level	Bachelor	Junior high school	High school
No religious belief	97.3%	70.5%	91.1%
Have religious belief	2.7%	29.5%	8.9%

analysis of variance (ANOVA) and regression analysis technique was used for relevant analysis between constructivism and environmental consciousness. Comparison studies between rural residents, urban residents, and migrants are throughout the research.

First, we test whether the NEP scores of these three groups are the same. This is by using the one-way ANOVA test since the dependent variable is an interval variable and the independent variable is a categorical variable. We used the significance and eta^2 to analyze the eco-consciousness level difference between the groups. The significance is also commonly known as *P*-value, and it is the probability that the hypothesis, for example, the NEP of these three groups are different in this case and is found to be rejected. Therefore, the lower the significance, the larger is the probability that this test is true. Here, we consider the significance $<.05$ as the criteria to confirm the difference. However, only the significance is not enough since it only confirms the difference but lacks the size of the effect. To include this information, eta^2 is used to measure the effect size. It is a value between zero and one, and could indicate the proportion of total variability explained by the independent variable. A larger effect size means a larger difference between the mean values. The eta$^2 = 0.01$ is considered as a small effect size, eta$^2 = 0.06$ a medium effect size, and eta$^2 = 0.14$ a large effect size (Stevens, 1999).

Table 4.3 shows the analysis result of the one-way ANOVA test. From the result, we can see the difference is significant with a large effect size. A post hoc Tukey HSD test is also performed to check the significant between every two groups. Table 4.3 shows between every two groups, the difference is all significant. It is confirmed that the three groups have a large difference in NEP.

Table 4.3 Analysis result of the NEP depends on the three groups of residents.
ANOVA table

		Sum of squares	df	Mean square	*F*	Sig.
NEP[a] three groups	Between groups (combined)	7934.904	2	3967.452	73.957	0.000
	Within groups	43130.861	804	53.645		
	Total	51065.765	806			

Measures of association

	Eta	Eta squared
NEP[a] three groups	.394	.155

Multiple comparisons
NEP
Tukey HSD

(I) three groups	(J) three groups	Mean difference (I − J)	Std. error	Sig.	95% Confidence interval	
					Lower bound	Upper bound
Rural	Migrants	− 6.74591[a]	.67610	.000	− 8.3334	− 5.1584
	Urban	− 8.50155[a]	.72353	.000	− 10.2004	− 6.8027
Migrants	Rural	6.74591[a]	.67610	.000	5.1584	8.3334
	Urban	− 1.75564[a]	.59116	.009	− 3.1437	− .3676
Urban	Rural	8.50155[a]	.72353	.000	6.8027	10.2004
	Migrants	1.75564[a]	.59116	.009	.3676	3.1437

[a]The mean difference is significant at the 0.05 level.

After we confirmed that the differences between groups are significant, we plot the histogram of the NEP for each group of residents, as shown in Fig. 4.3. Here, we use the traditional NEP scale, which ranges from 15 to 75 points. This figure shows: first, the previous research that the level of ecological consciousness of the Urban Resident Group is higher than the Rural Resident Group is verified. Second, the hypothesis that the environmental consciousness of migration is between the Urban Resident Group and the Rural Resident Group is verified. The mean value of NEP of the Migrant Group is 56.45, which is between 49.70 and 58.20 for the Rural Resident Group and the Urban Resident Group, respectively. Third, the environmental consciousness level of the Migrant Group and the Urban Resident Group are close. The residents with the NEP scores within 50−59 are very similar in the Migrant Group and the Urban Resident Group. However, for NEP above 65 points, the Urban

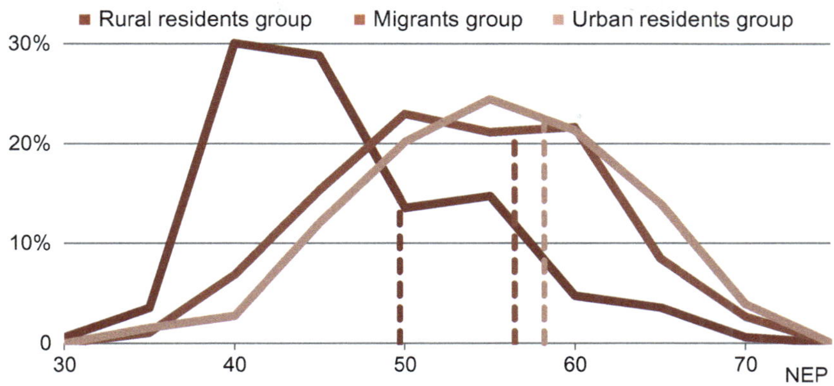

Figure 4.3 Histogram of the NEP for the three groups of residents.

Resident Groups has a slightly higher ratio, which is 17.8% compared with 11.1% in the Migrant Group.

From the previous discussions, a significant difference in the NEP can be found between the three groups of residents. Two questions in the first section of this chapter have been answered, that is, the environmental consciousness of urban residents and immigrants in cities at the same time and location are different, and rural residents and migrants have differences in their environmental consciousness. After these questions have been answered, one may ask, what is the reason for this difference?

To find out whether the socio-demographic variables cause the differences, linear regression is used to analyze the questionnaires with socio-demographic variables added as control variables. Here, the linear regression is chosen based on the control and dependent variables, that is, the socio-demographic variables and the NEP are interval variables. Dummy variables, that is, sex and religion are used with value one means male and have a religion, respectively. The independent variable, that is, the three groups of migrants is converted to dummy variables to compare between every two groups. We used the significance, *B*, and beta values to analyze the socio-demographic variables between the three groups of residents. The significance is introduced in this section. Here, we consider three levels of significance and mark them with stars, as listed in the bottom of Table 4.4. *B* value is the regression coefficient. It shows how much the dependent variables changes with the independent variables. A negative *B* value shows the dependent and independent variables change in different directions, for example, when the independent variable increases, the

Table 4.4 Analysis of socio-demographic variables and the NEP.

	Migrants—Urban	**Rural—Migrants**	**Rural—Urban**
	***B* (beta)**	***B* (beta)**	***B* (Beta)**
Sex	.128 (.009)	−.214 (−.013)	−.550 (−.033)
Age	−.044 (−.076)*	−.052 (−.098)**	−.106 (−.163)***
Education	1.587 (.267)***	1.762 (.325)***	1.414 (.265)***
Religion	−.110 (−.006)	−.207 (−.017)	.009 (.001)
Migrants—Urban	.843 (.056)		
Rural—Migrants		3.908 (.219)***	
Rural—Urban			5.167 (.292)***
Constant	51.142***	47.249***	50.616***
F (df)	13.207***	36.900***	39.485***
Adjusted R^2	0.088	.260	.330

***: $P < .01$; **$P < .05$; * $P < .1$.

dependent variable decreases. The beta value is the standardized value of the *B* value. It is normalized by the standard deviations of the variables.

Table 4.4 shows the regression analysis result of the three groups of residents. For the socio-demographic variables, age and education have significant effects on NEP of all the three groups and the sign of the *B* values are the same for all the three groups. People who are younger and accepted higher education tend to hold a higher level of environmental consciousness compared with people who are older or without a good education. This result was confirmed in the previous studies.

In addition, it is interesting to show that the Rural Resident Group still shows a significant difference with the other groups. The difference between the Rural Resident Group and the Urban Resident Group is consistent with previous research and our previous discussions. However, when comparing the Migrant Group and the Urban Resident Group with the control variables, the result shows the difference is due to the different distribution of the education level of these two groups since the education is significant; however, the Migrants−Urban dummy variable is not. The significance of the age cannot explain this difference since Table 4.2 shows the age is younger for the Migrant Group and the negative *B* value in Table 4.4 shows a younger age should have higher NEP. Based on the age, the Migrant Group thus should have a higher NEP than the Urban Resident Group which is different from the result shown in Table 4.4. In conclusion, the difference between the Migrant Group and the Urban Resident Group can be explained by the higher education level in average in the Urban Resident Group.

Table 4.4 also shows a significant effect between the Rural Resident Group and the Migrant Group. Note that the migrants are originally rural residents when they are young. They are a group of people who decide to move to large cities to change their destiny. They have a significant increase in the environmental consciousness compared with the residents who still live in rural areas. As our research design is based on the rapid urbanization scenario of China, the migrants may be considered as an important group of residents which is at the frontier of the changing in the environmental consciousness of China. Therefore, the paradox at the beginning of this chapter is solved and the hypothesis is verified: Migration is one important reason for the development of the environmental consciousness of China.

Can this conclusion be deeply explained? We have found the reason for the NEP difference between the Migrant Group and the Urban Resident Group. However, why the environmental consciousness of migrants has a rapid increase and have much higher NEP scores than rural residents? After reviewing previous theories, the next chapter will try to solve these questions and find out what drives the environmental consciousness based on data analysis.

CHAPTER FIVE

What drives environmental consciousness?

As discussed in the last chapter, find out the factors affecting the level of environmental consciousness can help us understand the reasons for the differences in environmental consciousness among the three groups of residents and lay the foundation for further environmental behavior research. In addition, aiming at the characteristics of migrants which is a group of residents with time and location transformed, the fundamental reasons for the change of the environmental consciousness may be found.

In the previous research, many tentative and meaningful theoretical explanations on the formation and changing of the environmental consciousness were carried out, for example, pollution-driven theory, constructivism theory, and postmaterialism theory. In this chapter, the factors of these theoretical explanations are measured in the questionnaires and the results of the measurement are analyzed with the level of environmental consciousness. This analysis may help us understand the real driving factor among the three theoretical explanations.

5.1 Pollution drive model

5.1.1 Theory review

From the objective angle, environmental consciousness may directly be affected by pollution. Therefore, the pollution-driven theory becomes one of the most important theories for explaining the level of ecological consciousness. Rapid industrialization and urbanization go hand by hand in China. They bring not only economic blooming but also environmental pollution. When pollution gets more serious, it awakens the environmental consciousness and then setting up eco-consciousness.

Environmental Consciousness in China.
DOI: https://doi.org/10.1016/B978-0-08-100388-6.00005-2

In previous research, Yan believes that environmental consciousness is based on the existence of society. In the modern sense, it was emerged synchronized with environmental pollutions (Yan, 2008). Moreover, Tong said that if environmental pollution threats the safety of people's life, awareness must be raised. The preference for environmental policies will be affected too. In other words, the local residents live at the place where serious pollution or specific contamination happened may have a higher level of environmental consciousness. Thus, the degree of pollution is the key that determines the environmental consciousness and policy preferences (Tong, 2002). Furthermore, Stern et al. stated that the type of environmental consciousness driven by pollution is based on the consideration of the individual's self-interest. In this case, the environmental protection activities are passive responses which cannot last long (Stern, Dietz, & Karlof, 1993). In addition, Zhou declared people who participated in environmental protection because of their concerns on specific environmental issues, rather than the overall consciousness of the natural environment. Therefore, one might take the initiative to participate in environmental organizations or make environmental actions by the pollution drive, but they tend not to give too much attention to environmental problems far away from them (Zhou, 2008).

5.1.2 Description of the questionnaires

Based on the pollution theory, the following questions are designed for measuring the perception of environmental pollution of the residents. In the first question, we directly measured the experience of pollution suffered, as listed in Table 5.1. According to the pollution-driven theory, people who have suffered environmental pollution may have a higher level of environmental consciousness.

In the second to the fifth question, the level of environment condition felt by the residents is measured. The four questions represent four environment dimensions which are from the Quality of Life scale (QOL) (World Health Organization, 1991), as listed in Table 5.1. According to the pollution-driven theory, people who feel a lower level of environment condition may have a higher level of environmental consciousness.

The last two questions are designed based on a new hypothesis: the environmental consciousness of migrants may be affected by the relative pollution, as listed in Table 5.1. The value of relative pollution only can

Table 5.1 The list of the questionnaires for the pollution-driven theory.

No.	Questionnaires	Scale
1	Have you or your family been exposed to any environmental pollution or have any health or economic loss due to environmental pollution within the past year?	0. No 1. Yes
2	How healthy is your physical environment?	The scale is using 5-point Likert scale with five levels: strongly disagree, disagree, neither agree nor disagree, agree, and strongly agree. Questions 2, 4, 5 are positive dimensions, i.e., respondents answered from strongly disagree to strongly agree were recoded as 1, 2, 3, 4, 5. The third question is a negative dimension, i.e., respondents answered from strongly disagree to strongly agree were recoded as 5, 4, 3, 2, 1. The points of the four questions are summed to get a value represent for the level of the surrounding environment conditions the responders feel. A larger value represents a more satisfied feeling.
3	How concerned are you with the noise in the area you live in?	
4	How satisfied are you with your physical environment (e.g., pollution, climate, noise, attractiveness)?	
5	How satisfied are you with the climate of the place where you live?	
6	Between 0 and 100, please rate your physical environment now (100 means the best).	These questions for only for migrants to evaluate the feeling changes of the physical environment between the city and rural area. The value of question 6 minus the value of question 7 gives the relative pollution value. More negative value means larger decrease of the physical environment feeling after migrate.
7	Between 0 and 100, please rate your physical environment you were living in the rural area (100 means the best).	

be tested with the Migrant Group and positive or negative score means better or worse environmental condition feelings after migrate since the data in this survey is filtered with only the migrants from the rural areas to the city. According to pollution-driven theory, a more negative value may lead to higher environmental consciousness.

5.1.3 Verify the pollution-driven theory

Table 5.2 shows the distributions of the above questions. It shows that many people have pollution experience in the city than in rural areas. It also shows that urban residents and migrants have a worse feeling of the surrounding physical environment than the rural residents. In addition, the migrants feel a decrease in the physical environmental condition after migrating on average; however, there are still many migrants who feel an increase. With a reminder that the rural residents have a lower environmental consciousness than the migrants and urban residents as discussed in Chapter 3: From Movements to Measurement, the environmental consciousness clearly shows a correlation with the pollution-driven theory except for the relative pollution. However, many factors are different between rural areas and the city, and the location should be controlled in our following analysis to check the pollution-driven theory for each group of residents.

A regression analysis was conducted to analyze the correlations between the pollution-driven theory and NEP. Linear regression is used since the independent variables are interval variables and a dummy variable, and the dependent variable is an interval variable. Table 5.3 lists the regression result. The table shows the pollution experience is a significant factor for all three groups of residents; however, the physical environment is not a significant factor for any of the three groups of residents. Comparing these two dimensions, the pollution experience mainly focuses on the loss, either health or economic, due to the pollutions. In contrast, the physical environment is only based on the environmental conditions the responders felt. This result shows that if environmental pollution causing any loss to the public, awareness is raised. Compared to the direct loss, indirect deteriorate of the surrounding environment does not have a significant effect. Moreover, the pollution experience has a smaller

Table 5.2 Distribution of answers of the questions for the pollution-driven theory.

	Questions		Rural	Migrants	Urban
1	Pollution experience	Mean:	0.234	0.336	0.414
2–5	Physical environment	Mean:	12.89	10.68	10.90
		Std.:	2.19	2.55	2.46
		N:	176	381	259
6–7	Relative pollution	Mean:		− 3.64	
		Std.:		13.06	
		N:		359	

Table 5.3 The regression result of the correlations between the pollution-driven theory and NEP.

Pollution-driven theory and NEP	Rural residents	Migrants	Urban residents
	B (beta)	*B* (beta)	*B* (beta)
Pollution experience	4.921 (.283)***	2.311 (.156)**	3.622 (.240)***
Physical environment	.246 (.075)	.179 (.064)	− .149 (−.049)
Relative pollution		− .241 (−.438)***	
Constant	55.250***	57.465***	65.568***
F (df)	7.100 (2)***	34.329 (3)***	7.814 (2)***
Adjusted R^2	.068	.222	.051

Note: ***$P < .01$; **$P < .05$; *$P < .1$.

influence on the migrants than the other two groups of residents. This is true for both control and not control the relative pollution. In addition, for Migrants Group, relative pollution affects environmental consciousness very significantly. Comparing the beta value with the pollution experience dimension, the relative pollution has a much larger influence on the NEP, showing that the change in the environmental conditions play a very important role for the migrants in the pollution-driven theory.

In conclusion, the pollution-driven theory has been verified in all three groups of residents. The direct loss due to pollution and relative pollution are the main factors driven by environmental consciousness. For the migrants, the experience of living in two places with a decrease in the environmental conditions may thus be a key factor for the rapid increase of the environmental consciousness of the migrants.

5.2 Constructivism model

5.2.1 Theory review

The last section discussed the driving model in the objective angle. If we are trying to understand the change of the environmental consciousness from another angle, it is the subjective angle. In 1995, Hannigan pointed out "in the early 1990s, environmental scholars were still by and large committed to a strong 'realist' position, whereby the gravity, configuration, and causes of ecological damage and destruction were thought to be obvious. By contrast, many researchers recognized that environmental risks and knowledge were by no means self-evident, but rather the product of social definition and construction" (Hannigan, 1995). Hannigan also pointed out that "Environmental problems do not materialize by themselves; rather, they must be 'constructed' by individuals or organizations who define pollution or some other objective condition as worrisome and seek to do something about it" (Hannigan, 1995).

This view is related to the agenda-setting theory. A well-known dictum by political scientist Bernard Cohen claimed, "The press may not be successful much of the time in telling people what to think, but it is stunningly successful in telling its readers what to think about" (Cohen, 1963). In the United States, important issues are put on the agenda by mass media; the typical example is environmental protection in the 1970s. Agenda-setting theory shared the same idea with Anthony Downs's "issue-attention cycle"—the rise and fall of an issue on the public agenda (Downs, 1972).

However, Dunlap points out that "Downs' issue-attention cycle is not an inexorable process." He stated that public support for environmental protection has gone through its infancy, arousal, climax, and decline, but is still in sight. Even after the government withdrew its attention from the environment, mass environmental consciousness made the 1980s revive after some recession (Dunlap, 1995).

5.2.2 Description of the questionnaires

Based on Hannigan's opinion, science and mass media are two important dimensions of constructivism. In addition, previous studies have shown that knowledge of the environment is an important intermediary that leads to environmental consciousness (Hannigan, 1995). A higher level of environmental knowledge often results in more concern of environmental (Arcury, Scollay, & Johnson, 1987; Hayes, 2001). Therefore, the constructivism theory in this chapter will be operating into three dimensions: scientism, opportunities to obtain information, and environmental knowledge.

Scientism is derived from Faith in Science. Previous researches including people's beliefs during industrialization have specialized measurement of scientism. Science and Technology Scale (Baxter, 1988) and the Dominant Social Paradigm (Dunlap & Van Liere, 1984) are two classic models for this research. Combining two scales to one is not our initiative but the results from a careful selection. Carton (Carton, 2010) verified that these two scales have common factor (eigenvalue) >.7, with acceptable reliability ($\alpha = .75$). He believes they are well combined because they all are related to the ability to deal with human problems with science and technology. The questions of Scientism are listed as Questions 1–4 in Table 5.4. The residents with a higher score of Faith in Science and Technology are expected to have a higher environmental consciousness.

The questions in the Opportunities of Obtain Information dimension are selected from World Health Organization Quality of Life, 100 items (WHOQOL-100) (World Health Organization, 1991), from the World Health Organization (WHO). The environmental dimension in WHOQOL-100 composed of the items including Opportunities to Obtain Information. Since reliability and validity of WHOQOL-100 have been verified and WHOQOL-100 have already verified in China, our questionnaire also using the Opportunities of Obtain Information scale, as shown as Questions 5–8 in Table 5.4, and a higher score may lead to higher environmental consciousness.

The Environmental Knowledge score is adapted from the environmental part of the Chinese General Social Survey (CGSS) 2010. The first half of the questions are environmental knowledge in daily life, and the rest of the questions are environmental knowledge in popular science. The questions are listed as Questions 9–20 in Table 5.4 and a higher score is predicted to have a higher environmental consciousness.

5.2.3 Apply the constructivism model in China

Table 5.5 shows the distribution of answers to the questions of the constructivism model. In the table, the Faith in Science and Technology is almost the same for the three groups of residents. In the other two dimensions, the rural residents show smaller opportunities to obtain information and less environmental knowledge than the migrants and urban residents. Considering the lower environmental consciousness of the rural residents than the migrants and urban residents, it seems these two dimensions are correlated with the environmental consciousness. To better understand the correlations within each group of residents, regression analysis with the three groups of residents controlled is then performed.

A regression analysis was conducted to analyze the correlations between the constructivism theory and NEP. Linear regression is used since the independent variables are interval variables, and the dependent variable is an interval variable. Table 5.6 shows the regression analysis

Table 5.4 The list of questionnaires of the constructivism model.

No.	Questions	Scale
1	Science and technology have improved our quality of life.	Using 5-point Likert scale: strongly disagree, disagree, neither agree nor disagree, agree, and strongly agree. Questions 1 and 3 (2 and 4) are positive (negative) dimensions, i.e., respondents answered from strongly disagree to strongly agree were recoded as 1, 2, 3, 4, 5 (5, 4, 3, 2, 1), respectively. The score for scientism is the sum of these four questions.
2	Complex technologies will always be risky because of the chance of human error.	
3	Most problems can be solved by applying more and better technology.	
4	We cannot keep counting on science and technology to solve humankind's problems.	
5	How available to you is the information that you need in your day-to-day life?	Using 5-point Likert scale. For Questions 5 and 6, not at all, a little, moderately, mostly, completely corresponding to score 1, 2, 3, 4, 5. For Questions 7 and 8, very dissatisfied, dissatisfied, neither satisfied nor dissatisfied, satisfied, very satisfied corresponding to score 1, 2, 3, 4, 5. The score for Opportunities of Obtain Information is the sum of these four questions.
6	To what extent do you have opportunities for acquiring the information that you feel you need?	
7	How satisfied are you with your opportunities for acquiring new skills?	
8	How satisfied are you with your opportunities to learn new information?	
9	Vehicle exhaust does not harm human health.	Respondents is asked to choose one of the three choices: correct, incorrect, do not know. The correct choices for Questions 9, 11, 13, 15, and 17 is "correct," and the other questions is "incorrect." Each correct selection worth one point. Incorrect selections or choose "do not know" has no credit. The environmental knowledge score is the sum of the score of all these questions.
10	Abuse fertilizers will cause soil acidification, and resulting in environment damage.	
11	Using powder detergent containing phosphate will not cause water pollution.	
12	Fluoride emissions from refrigerators will be the factors that destroy the ozone layer.	
13	There is no relation between acid rains and coal firing.	
14	Creatures are interdependent with each other. The disappearance of one species will have a ripple effect.	

(*Continued*)

Table 5.4 (Continued)

No.	Questions	Scale
15	In air quality reports in China, the 3rd level is better than the 1st level.	
16	Single species of tree is more likely have diseases and pests.	
17	In water pollution reports in China, V (5th) level water shows better quality than I (1st) level.	
18	Increase carbon dioxide in the atmosphere is one of the factors for global warming.	
19	Material for composting including human and animal's feces.	
20	The pesticide will harm people through spray and mucosa contacted.	

Table 5.5 Distribution of answers of the questions of the constructivism model.

	Questions		Rural	Migrants	Urban
1–4	Faith in Science and Technology	Mean:	12.42	12.48	12.27
		Std.:	1.69	2.01	1.95
		N:	172	378	258
5–8	Opportunities to Obtain Information	Mean:	10.77	11.81	12.07
		Std.:	2.05	2.44	2.38
		N:	176	381	259
9–20	Environmental Knowledge	Mean:	6.82	8.97	8.94
		Std.:	2.91	2.12	2.02
		N:	169	359	246

results of the three dimensions of constructivism for three groups of residents. In the table, the result of the three dimensions is different. Hannigan believes that science and mass media are two important dimensions of constructivism. For the science dimension, the faith in science and technology is significantly related to the environmental consciousness which obtained the data support from rural residents and migrants. These two groups of residents both have a higher environmental consciousness when the faith in science and technology increases. However, previous researches conclude that a person's faith in science reflects the extent to which they believe that scientists can solve the problems of humankind

Table 5.6 Regression analysis for the three groups of residents.

Constructivism model and NEP	**Rural residents**	**Migrants**	**Urban residents**
	***B* (beta)**	***B* (beta)**	***B* (Beta)**
Faith in Science and Technology	1.230 (.279)***	. 521 (.140)***	.320 (.084)
Opportunities to Obtain Information	.120 (.033)	–.098 (–.032)	–.206 (–.067)
Knowledge of the Environment	.187 (.074)	.396 (.114)**	.042 (.285)***
Constant	31.915***	47.527***	47.592***
F (df)	5.176 (3)***	3.949 (3)*	7.297 (3)***
Adjusted R^2	.072	.024	.072

Note: ***$P < .01$; **$P < .05$; *$P < .1$.

(Dunlap & Van Liere, 1984), then they recognize environmental problem would be handled by science and technology that people not necessarily worried about it.

The difference between the prediction and Table 5.6 indicates that the prediction may not be the dominant effect in China. Indeed, the education level is low in rural areas in China, and people with lower faith in science and technology usually have a lower education level. They may know very little about the relationship between environments and humans. This effect is dominant for the migrants and rural residents, however, for urban residents, most of them may have enough education to start questioning the science and technology and learn the limitations of them. Less faith thus forms as the aforementioned research predicted. With these two effects, the faith in science and technology has no significance with the environmental consciousness. This two-stage effect also causes the mean faith in science and technology between these three groups of residents almost the same although the education levels are very different, as shown in Table 5.5.

For the second important dimension of constructivism pointed out by Hannigan, that is, mass media, the opportunities to obtain information are used to resemble this effect (Hannigan, 1995). However, it does not show a significant relationship with the environmental consciousness for all the three groups of residents, as shown in Table 5.6. It may be due to Hannigan's theory not available in China or the opportunities to obtain information do not fully resemble the mass media effect. The mass media is considered broadcast and residents receive information passively with very limited methods to select. In contrast, nowadays, with the widespread of the internet, people start to actively obtain the information they are interested by searching with keywords, which may not be related to environments. With this in mind, the non-significance may be understood.

To better analyze the effect of the mass media, an additional dimension, that is, knowledge of the environment is added to measure the environmental knowledge of the residents really obtained. Table 5.6 shows the knowledge of the environment is significantly related to the environmental consciousness for both urban residents and migrants. Positive *B* value indicates that residents with more knowledge of the environment tend to have higher environmental consciousness, which is consistent with our prediction.

The standardized slope (beta) in three groups illustrates a continuous trend. In detail, the slope of the level of environmental knowledge score with the NEP in the three groups is in a gradual rising trend and the slope of scientism score with the NEP level in the three groups is in a gradual declining trend. Although the opportunities to obtain information do not have a significant relationship with the NEP, a trend of the beta value is also found by comparison of three groups. Since we selected the samples in this analysis, rural residents, migrants from rural to urban, and urban residents could be recognized in one linked spectrum. These trends give the expression of the transformation from rural to urban.

In summary, we verified the constructivism model which drives the NEP. Both the scientism and the knowledge of the environment have positive correlations with the NEP. The opportunities to obtain information are not correlated with the NEP since it may not be well represented by the mass media in the constructivism.

5.3 Postmaterial model

5.3.1 Theory review

In the last two sections, the objective angle and subjective angle are discussed. Besides these two angles, the economy may be another important factor in environmental consciousness. In this section, we try to discuss from the economic angle, and trying to find whether "postmaterialism" will drive the environmental consciousness. "Postmaterialism" was declared by Ronald F. Inglehart in 1977. Inglehart addressed the value changes from materialism to postmaterialism in the advanced industrial society. Materialists believe in economic growth and material security. On the contrary, postmaterialists put materialism behind nonmaterial values such as environmental protection, freedom of speech, and gender equality (Inglehart, 1977). Environmental sociologists agreed with this opinion and stated that people were concerned about environmental quality only if the basic material needs have been satisfied (van Liere & Dunlap, 1980).

China's environmental sociologists also contributed their opinions on the postmaterialism approach. Hong pointed out that people who live in higher economic development areas are more concerned about environmental protection and hold stronger environmental consciousness than people living in less developed areas. In China, Hong found a very positive relationship between postmaterialism value and environmental consciousness. He stated that the environmental consciousness level rises hand by hand with the growing economy and better living standards, in general (Hong, 2005).

Tong believes that the economic level has an effective power to shift environmental consciousness and the priority of environmental policy. She said, when society developed to the phase of postmaterialism, with the food suffices and wealth safety, people's attention change from the amount of income to the quality of life. The environmental problem is one of the most important parts of life quality. Therefore, environmental consciousness will appear (Tong, 2002).

However, Hong and Tong both agreed that China has not joint in the advanced industrial society yet, and postmaterialism is not the mainstream value. Tong even got contradictory findings against the postmaterialism theory based on her research for six cities (Beijing, Lanzhou, Guangzhou, Shenyang, Shanghai, and Chengdu) in China. Although values in five cities match the interpretation of the postmaterialism theory, fewer people in Shanghai supported the environmental protection policy which might slow down economic development. As a matter of fact, Shanghai sits on the top list of economic standards in China. This contradictory finding was not the only evidence that challenged the postmaterialism theory, in fact, the theory cannot well-explain the growth of environmental consciousness and the increasing environmental movement in the lower-income area.

5.3.2 Description of the questionnaires

In order to quantify the materialism, we use four dimensions to measure the quality of living of the respondents. The four dimensions adopted from the WHOQOL are housing conditions, financial conditions, medical conditions and social security conditions, that is, traffic conditions. Table 5.7 listed the 16 questions for the postmaterial measuring. Every four questions are in one dimension. A higher value of the four conditions represents a better living condition, and under the postmaterialist theory, it is expected to have more consciousness on the nonmaterial values, for example, environments.

5.3.3 Apply the postmaterial model in China

Table 5.8 shows the distribution of the answers to the questions of the postmaterial model. In the table, the score of urban residents is higher than the residents in the other two groups. This result seems consistent with the postmaterial theory, and it is the same as the previous conclusions in the study of public environmental awareness in China. Hong pointed out that in areas with high levels of economic development, people tend to pay great attention to environmental protection, attach importance to the coordinated development of people and nature, and have a high level of environmental awareness (Hong, 2005). In general, with the increase in public income and the improvement of living standards, the public will increasingly care about environmental protection (Yan, 2007).

This promising result may support the postmaterial theory; however, to confirm whether they are directly related or not, we should use the three groups of residents as a control variable, that is, confirm within each group of residents, the people with higher material conditions will have higher environmental consciousness. A regression analysis was conducted to analyze the correlations between the postmaterial theory and NEP. Linear regression is used since the independent variables are interval variables, and the dependent variable is an interval variable, as shown in Table 5.9. According to the analysis result, the theoretical assumptions of postmaterialism have not been verified since the table shows no significance except the traffic conditions in the Urban Resident Group (Sig $<$ .05). To understand this result, it is worth noting that in Table 5.8, most of the mean values are close to the score of the "neither agree nor disagree" option, which is 12 for four questions. Two dimensions, that is, the financial conditions and the medical and social security conditions

Table 5.7 The list of questionnaires for the postmaterial model.

No.	Questions	Scale
1	How comfortable is the place where you live?	Using 5-point Likert scale: strongly disagree, disagree, neither agree nor disagree, agree, and strongly agree. All the four questions are positive scales, i.e., respondents answered from strongly disagree to strongly agree were recoded as 1, 2, 3, 4, 5, respectively. The score for the housing conditions is the sum of these four questions.
2	To what degree does the quality of your home meet your needs?	
3	How satisfied are you with the conditions of your living place?	
4	How much do you like it where you live?	
5	Have you enough money to meet your needs?	Using 5-point Likert scale: strongly disagree, disagree, neither agree nor disagree, agree, and strongly agree. Questions 5 and 7 (6 and 8) are positive (negative) scales, i.e., respondents answered from strongly disagree to strongly agree were recoded as 1, 2, 3, 4, 5 (5, 4, 3, 2, 1), respectively. The score for the financial conditions is the sum of these four questions.
6	Do you have financial difficulties?	
7	How satisfied are you with your financial situation?	
8	How much do you worry about money?	
9	How easily are you able to get good medical care?	Using 5-point Likert scale: strongly disagree, disagree, neither agree nor disagree, agree, and strongly agree. All the four questions are positive scales, i.e., respondents answered from strongly disagree to strongly agree were recoded as 1, 2, 3, 4, 5, respectively. The score for the medical conditions and social security conditions is the sum of these four questions.
10	How would you rate the quality of social services available to you?	
11	How satisfied are you with your access to health services?	
12	How satisfied are you with the social care services?	
13	To what extent do you have adequate means of transport?	Using 5-point Likert scale: strongly disagree, disagree, neither agree nor disagree, agree, and strongly agree. Questions 13 and 15 (14 and 16) are positive (negative) scales, i.e., respondents answered from strongly disagree to strongly agree were recoded as 1, 2, 3, 4, 5 (5, 4, 3, 2, 1), respectively. The score for the traffic conditions is the sum of these four questions.
14	To what extent do you have problems with transport?	
15	How satisfied are you with your transport?	
16	How much do difficulties with transport restrict your life?	

Table 5.8 Distribution of answers of the questions of the postmaterial model.

	Questions		Rural	Migrants	Urban
1–4	Housing conditions	Mean:	12.19	12.27	12.86
		Std.:	2.32	3.29	3.16
		N:	176	381	259
5–8	Financial conditions	Mean:	11.91	11.83	12.22
		Std.:	2.87	3.81	3.69
		N:	176	381	259
9–12	Medical and social security conditions	Mean:	10.98	10.73	11.20
		Std.:	2.38	2.62	2.70
		N:	176	381	259
13–16	Traffic conditions	Mean:	12.56	12.81	13.37
		Std.:	2.14	3.33	3.38
		N:	176	381	259

Table 5.9 Regression analysis of the postmaterial model for the three groups of residents.

Postmaterial and NEP	Rural	Migrant	Urban
	B (beta)	*B* (beta)	*B* (beta)
Housing conditions	− .249 (−.078)	.157 (.071)	− .034 (−.015)
Financial conditions	.318 (.125)	.162 (.085)	.221 (.109)
Medical and social security conditions	− .404 (−.133)	− .067 (−.172)	− .018 (−.006)
Traffic conditions	− .342 (−.101)	− .203 (−.093)	− .405 (−.185)**
Constant	57.675***	55.930***	61.569***
F (df)	1.701 (4)	1.151 (4)	1.923 (4)
Adjusted R^2	.016	.002	.014

Note: ***$P < .01$; **$P < .05$; *$P < .1$.

have a mean score less than the "neither agree nor disagree" option. It shows many respondents, including the urban respondents, are not satisfied with their material conditions, and thus may not have a postmaterial value. As a result, China may not enter the postmaterial stage at the time of the survey, so it may be the reason that the correlation trend between postmaterial score and environmental consciousness is not obvious.

In addition, the only significant relation, that is, the transportation condition for the Urban Resident Group has a negative *B* value, which means the lower the transportation condition score, the higher is the residents' environmental consciousness. This relationship does not support the postmaterial theory since according to the postmaterial theory, the better traffic condition should focus more on the nonmaterial values and have a higher environmental consciousness. One possible reason may be a worse traffic condition becomes a type of pollution and drives the environmental consciousness based on the pollution drive theory. This pollution-driven result only shows for the urban residents, but not the other two groups of residents. To understand this, take into account that in the survey, urban residents and rural residents are concerned about different environments. For example, urban residents pay more attention to traffic congestion, noise pollution, and air pollution, while rural residents are more concerned about drinking water safety and reduced cultivated land. As a result, the traffic condition should only be related to the group of residents who pay more attention to it, that is, urban residents.

Overall, our data show the postmaterial theory does not apply to China for now since China may not enter the postmaterial stage. The only exception that traffic conditions related to the environmental consciousness of the urban residents may be driven by pollution.

5.4 Conclusion

This chapter tried to answer the questions we raised in the last chapter, that is, how to understand the environmental consciousness change in China? To answer the question, three major factors, that is, pollution theory, constructivism theory, and postmaterial theory are considered in order to find the driven mechanism. In addition, we use the three groups of residents as a control variable to isolate the effect due to the location. The regression analysis result shows that the rapid increase of the environmental consciousness in China is mainly related to the pollution theory and constructivism theory. In addition to understanding the mechanism of the rapid environmental consciousness increase, this mechanism may also explain the reason for the even faster increase of the environmental consciousness of the migrants. As the migrants move from the rural areas to the city, the pollution they conducted and information or knowledge they can access quickly changes to a level near the urban residents. Their environmental consciousness may thus rapidly increase due to these changes.

From the first chapter to now, we discussed the environmental issues, movements, measurements of the environmental consciousness, and the driving mechanism of the environmental consciousness. With the environmental consciousness, people are trying to protect the environment, which we will discuss in the next chapter.

CHAPTER SIX

Does consciousness convert to behavior?

In Chapter 2, Environmental Issues to Protection: The US versus China of this book, we have discussed most environmental issues that occur in modern times after the industrial revolution was caused by human activities, that is, the secondary environmental issues. Therefore, understanding human–environmental activities are always the final goal for most environmental researchers. They have been trying to find a way to reduce the secondary environmental issues by understanding and guiding the environmental behavior of the public.

The "environmental behavior," as Poortinga et al. called it (Poortinga, Steg, & Vlek, 2004), has been used by many previous researchers, but it is also good to know that different researchers are using different terms when they refer to environmental behavior. One of the differences is that in the narrow sense of "environmental behavior," the term only identifies the positive activities related to the environment, such as recycling resources or protecting the environment. In contrast, in the general sense, "environmental behavior" could include negative environmental activity, like wasting energy. As the research is mainly related to how to protect the environment, previous researches usually focus only on positive behaviors, for example, Gatersleben et al. preferred "pro-environmental behavior" (Gatersleben, Steg, & Vlek, 2002) and Kaiser think "eco-behavior" is better (Kaiser, Wlfing, & Fuhrer, 1999). These different words are all for narrow senses. Furthermore, some other researchers would like to state the term with an even more narrow sense and expressive meaning. For example, Stern stated as "Environmentally Significant Behavior" (Stern, 2000), and Hines called it "responsible environmental behavior" (Hines, Hungerford, & Tomera, 1986). Although these terms are different in expression, they have the same connotation. They all emphasize the individual's participation to prevent or solve environmental problems.

In this book, we will use pro-environmental behavior in the first part of the discussions, which will include willingness, daily activities, and

Environmental Consciousness in China.
DOI: https://doi.org/10.1016/B978-0-08-100388-6.00006-4

strive to change. In the second part of our discussions, we will use the planned behavior as the basic model, which makes it "responsible environmental behavior."

6.1 Three levels in pro-environmental behaviors

How people conduct pro-environmental behavior in daily life is still a complex question. For example, the relationship between environmental consciousness and pro-environmental behavior is still being argued by many researchers.

During the debate, previous researchers found that it will be clearer if they organize pro-environmental behavior by different categories and fill them into different levels based on the difficulties. For example, the environmental behavior with the least difficulty is to claim pro-environmental behavior can be the "willingness to do." Although it is not the real activity, the mainstream considered the willingness as an important part of the environmental behavior. For example, in a typical definition by Stern, two dimensions are proposed, that is, "intention" and "impact" to define environmental behavior. "Intention" oriented definition emphasizes actors' environmental motivation or willingness and "impact" oriented definition emphasizes the impacts to the environment (Stern, 2000). Actually, the willingness has been found closely related to environmental consciousness (Schuhwerk & Lefkoff-Hagius, 1995; Vlosky, Ozanne, & Fontenot, 1999).

Moreover, people hold higher agreement on willingness to do rather than really doing the environmental activities. As a result, the next level with higher difficulty is daily protection activities. In addition, even if we focus on real environmental behavior, a big gap has shown on the achievements' level between personal activities and the interactivities with other people. In other words, personal daily protection activities are easier to achieve, but it is harder to conduct public environmental behavior, like political or legal behaviors. This is because public environmental behaviors require a competent capacity of residents, including environment and legal knowledge (Sawitri, Hadiyanto, & Hadi, 2015). Researchers in China also found the decline tend from personal environmental behavior to public environmental behavior. Ren claims there are different levels of environmental behavior. As the level rises, the difficulty of participate increases. She agreed with previous researchers, "Environmental movements request environmental knowledge and legal information, while personal subjectivity and political awareness are also needed" (Ren, 2002). Therefore, the frequency of public interactivities is low, and these public environmental behaviors to strive to change the surroundings is considered as the most difficult level of environmental behavior in this research.

Overall, based on the common findings from previous research, we considered our environmental behavior research in this book at three levels: willingness, daily protection activities, and strive to change the surroundings. These three dimensions cover three levels of environmental activities that include the link between the willingness and the impacts and the cooperation between personal activities and strive to change the public. In this chapter, these three dimensions are measured and analyzed.

6.2 Willingness

The first tested environmental behavior level is willingness. We split the willingness into two aspects, that is, willing to change and willing to pay. These two aspects measure the willingness of the responders to sacrifice their time, labor, or finance. The hypothesis is that people who have a higher environmental consciousness may have higher willingness-to-change behaviors in pro-environmental activities and willingness-to-pay (WTP) behaviors for the pro-environmental products.

To measure the willingness of changing behaviors to pro-environmental activities, a direct inquiry approach was used in our survey. People were asked, "are you willing to protect the environment by changing your routine habits, for example, saving water, preventing plastic bags, and resource recycling?" The 5-point Likert scale from "Not willing at all" to "very willing to" is used to measure the result. Respondents answered from "Not willing at all" to "very willing to" were recoded as 1, 2, 3, 4, and 5, respectively. A higher score represents a greater willingness to change the behavior.

Table 6.1 shows the distribution of the answers to the questions of the willingness. It shows most of the urban residents and migrants are willing to change their behavior and rural residents are generally neither willing

Table 6.1 Distribution of answers to the questions of the willingness.

Questions		**Rural**	**Migrants**	**Urban**
Willingness to change	Mean:	3.20	4.01	4.04
	Std.:	0.914	0.983	0.975
	N:	132	365	255
Willingness to pay	Mean:	2.63	1.95	1.93
	Std.:	1.040	0.771	0.754
	N:	176	381	259

Table 6.2 Environmental consciousness impacts on willingness-to-change.

	Rural	**Migrant**	**Urban**
NEP and willingness	***B* (beta)**	***B* (beta)**	***B* (beta)**
NEP	.034 (.271)***	.039 (.378)***	.045 (.412)***
Constant	1.520***	1.933***	1.534***
F (df)	10.160 (1)***	60.633 (1)***	51.586 (1)***
Adjusted R^2	.066	.140	.167

Note: ***: $P < .01$; **: $P < .05$; *: $P < .1$.

nor not willing. It, thus, shows the rural residents who have lower environmental consciousness as discussed in Chapter 4, From a Longitudinal Study to Cross-Sectional Study, have a significantly lower willingness to change their behavior than the urban residents and migrants. Therefore, it shows that environmental consciousness and willingness are correlated. To exclude the factors of different resident groups, an analysis within each group of residents is performed.

In the analysis, we use the willingness as the dependent variable and the NEP as the independent variable. Since these two variables are interval variables, a linear regression analysis is used for each group of residents, and so the three groups of residents are controlled. The regression result shows the willingness of change is significantly dependent on the NEP for all the three resident groups, as shown in Table 6.2. This result supports our hypothesis, that is, people who have held a higher level of environmental consciousness or NEP will hold stronger willingness to conduct pro-environmental behavior and protect the environment, for example, save water, prevent plastic bags, and resource recycling.

Moreover, the *B* values and beta values in the regression show a consistent increase from rural residents to urban residents. Thus, the impact of environmental consciousness on the willingness to change for the urban residents is larger than for the rural residents and for migrants is in between. This trend shows that although the mean value of the willingness of migrants is increased to a value close to the urban residents, the willingness to change of the migrants is less affected by the environmental consciousness than the urban residents.

Willingness to change measures the willingness of the responders to sacrifice their time or labor. In addition, previous researchers have investigated another topic, that is, the WTP, which measures the willingness of the responders to sacrifice their finance, in sustainable development areas. WTP is one of the important topics related to environmental consciousness that is considered by businessmen and policymakers. Previous researchers contributed many works on this topic and found out correlations have been found between environmental attitudes (using NEP scale) and general environmentally related WTP (Hansla, Gamble, Juliusson, & Gärling, 2008; Kang, Stein, Heo, & Lee, 2012; Kotchen & Reiling, 2000), and confirmed environmental consciousness could be used to estimate the WTP.

Our research kept this key question in the research questioner as well. The question was described, "Even costs are higher, I will use or purchase environmentally friendly products (organic vegetables, energy-saving utilization, etc.)" and asked responders to choose from 5 points scale including "never (occurred zero times)," "rarely (occurred <25% times)," "sometimes (occurred about halftime)," "most of the time (occurred more than 75% times)," and "always (occurred more than 95% times)." Respondents answered from "never" to "always" were recoded as 1, 2, 3, 4, and 5, respectively. A higher score represents a greater WTP.

Table 6.1 shows the distribution of the answers to the questions of the WTP. It shows almost 70% of responders in our research are not willing to purchase environmentally friendly products as a habit (most of the time or always) when the costs are higher. This result is much lower than the willingness-to-change indicates the residents are more willing to give up their labor or time than financial for the pro-environmental activities. These results repeated previous research in two ways: first, the willingness has been found to be the closest related to environmental consciousness. Second, but even people hold high agreement on willingness to change their personal behaviors, willingness to really purchase higher cost of environmentally friendly products is a totally different story. These results also reflect the purchase preference in China.

Table 6.1 also shows that rural residents are more willing to pay than urban residents and migrants. When considering the rural residents have a

Table 6.3 Environmental consciousness impacts on WTP.

NEP and willingness	Rural	Migrant	Urban
	B (beta)	*B* (beta)	*B* (beta)
NEP	− .025 (−.176)**	.005 (.044)	− .005 (−.047)
Constant	3.878***	1.689***	2.212***
F (df)	5.346 (1)**	.729 (1)	.568 (1)
Adjusted R^2	.025	− .001	− .002

Note: ***: $P < .01$; **: $P < .05$; *: $P < .1$.

lower environmental consciousness than the other two groups of residents, it shows a negative relationship between the environmental consciousness and the WTP, that is, the residents with a lower environmental consciousness are more willing to pay for the environmental friendly products. Similar to the willing to pay part, the three resident groups are controlled to show the relations within each resident group. In the analysis, we use the WTP as the dependent variable and the NEP as the independent variable. Since both these two variables are interval variables, a linear regression analysis is used for each group of residents, so the three groups of residents are controlled. The regression result shows the willingness of pay is not significantly dependent on the NEP only except the rural resident group, as shown in Table 6.3. The *B* value and the beta value also shows a negative relation between the NEP and the WTP.

To understand this relationship, we continued the question if the responders chose sometimes. An additional question is asked, "what situation will trigger you to buy environmentally friendly products?" The most answers we collected could be categorized into two situations. When it related to food, residents are willing to pay to avoid genetically modified organism (GMO) food. The other situation is for the items will save money in long time usage, for example, solar power panel. This result explains the residents are willing to pay when it benefits their health or finance in long terms but not directly related to the environmental consciousness.

6.3 Daily activities

After discussing the willingness to the pro-environmental activities, the next level of the environmental behavior is the daily activities. Similar to the willingness, an interesting question may be how environmental consciousness affects the frequency of conduct pro-environmental activities in daily life? Compared to the willingness, the pro-environmental activities were focused on responders' personal daily activities within the past 12 months. The daily actives, including reuse, reduce, recycle, and save energy and recourses, as listed in Table 6.4, is measured with the 10 protection behaviors that Hong et al. considered in their environmental behavior survey scale used in 2007 (Hong & Xiao, 2006).

For those 10 specific protection activities, responders will choose from 5 points scale including "never (occurred zero times)," "rarely (occurred <25% times)," "sometimes (occurred about halftime)," "most of the time (occurred more than 75% times)," and "always (occurred more than 95% times)." The answers from "never" to "always" were recoded as 1, 2, 3, 4, and 5, respectively. The total score is the sum of the scores of all these

Table 6.4 List of the daily activities questions and the distribution of the answers.

	Questions		Rural	Migrants	Urban
1	Save electricity (turn off lights, limit air-conditioning temperature in summer)	Mean:	3.09	4.37	4.37
		Std.:	1.40	0.53	0.53
2	Save energy (use public transportations, use solar panels)	Mean:	3.21	3.67	3.64
		Std.:	1.22	0.90	0.89
3	Save water (check water leaking regularly, use water-saving toilets, reduce bath time, etc.)	Mean:	3.21	3.98	3.91
		Std.:	1.18	0.72	0.76
4	Use gray water (for vegetables, flowers, and toilet flushing)	Mean:	2.90	2.11	2.05
		Std.:	1.24	0.81	0.72
5	Re-use (plastic bags, greenhouses, etc.)	Mean:	3.08	3.96	4.09
		Std.:	1.22	0.81	0.70
6	Use biogas, use waste heat from heating, etc.	Mean:	2.79	1.83	1.78
		Std.:	1.36	0.77	0.67
7	Recycle waste paper, plastic bottles, batteries	Mean:	3.20	3.81	3.98
		Std.:	1.25	0.80	0.74
8	Use own basket/shopping bag for daily purchase	Mean:	2.69	3.84	3.98
		Std.:	1.15	0.84	0.71
9	Avoid disposable products (wood chopsticks, paper towels, paper cups)	Mean:	2.58	2.39	2.37
		Std.:	1.06	0.97	0.94
10	Environment conservation (protect lawns, tree planting, etc.)	Mean:	2.51	1.49	1.46
		Std.:	0.66	0.62	0.56
	Total score	Mean:	29.20	31.44	31.63
		Std.:	8.10	2.80	2.34
		N:	176	381	259

10 questions, so the range of the total score is between 10 and 50. A higher score represents a higher frequency of pro-environmental daily activities.

Table 6.4 shows the distributions of the scores of these 10 questions and the total score. For most of the questions, the urban residents and migrants have a higher pro-environmental daily activity frequency than the rural residents, except Questions 4, 6, 9, and 10. Interestingly, the average score is generally low for these four questions compared with the other questions even for the rural residents. This may be due to the different lifestyles in rural areas and in cities, and lack of convenience. For example, residents in urban areas are using nature gas accessible in almost every apartment nowadays, and generating biogas or use waste heat from heating is not convenient and not necessary in urban areas. For the total score, urban residents who hold a much higher environmental conscious have a higher score than the rural residents, but the difference is not large.

To test our research hypothesis, that is, "people who hold high environmental conscious would have a high frequency of protection activities," we only focus on the effect by the environmental consciousness and have the three different groups of residents controlled. In this way, we test the relationship between the environmental consciousness and the pro-environmental daily activities within each group of residents, which have similar lifestyles. A regression analysis was conducted to analyze the correlations between the NEP and the total score of the pro-environmental daily activities. Here, linear regression is used since both the independent variable and the dependent variable are interval variables. Table 6.5 shows the results of the data analysis.

The result of the linear regression as shown in Table 6.5 did not support our hypothesis. People with a higher level of environmental consciousness did not find high-frequency protection actions. This result seems odd, but when we look closely at the result of the questionnaires

Table 6.5 Environmental consciousness impacts on the daily activities.

NEP and daily activities	Rural	Migrant	Urban
	B (beta)	*B* (beta)	*B* (beta)
NEP	.149 (.127)	.019 (.045)	− .023 (−.068)
Constant	39.423***	34.457***	34.888***
F (df)	2.747 (1)*	.781 (1)	1.197 (1)
Adjusted R^2	.010	.000	.010

Note: ***: $P<.01$; **: $P<.05$; *: $P<.1$.

and it can tell us why. As we can see, for example, item 4, "Use gray water (for vegetables, flowers, and toilet flush)," more than half of responders replied never or rarely. This may not be due to the lack of environmental consciousness, but due to lack of convenience. Gray water sounds a great idea to protect the environment, but in China, gray water collection is not in the system yet, and people rarely do it because it needs to put more time and work on it.

6.4 Strive to change the environment

The two easier levels in pro-environmental behaviors are not interacting with others. For the last level, it targets public level actions, which are the actions that strive to change the surrounding environment by interacting with others. In this level, we split the questions into two, that is, whether the respondents take any action, and which action did they take. In the following two subsections, we analyze them one by one.

6.4.1 Action or not

Environment deprivation is the pre-condition for people to strive to change their surroundings. In our research, environment deprivation is measured only for the respondents who have suffered environmental pollutions before, which has been discussed in Chapter 5, What Drives Environmental Consciousness?. For the respondents who choose "Yes" for the question "Have you ever suffered from environmental pollution?," they need to answer the consequence question: "Did you take any action to change this situation? Such as caution the polluters, go to the media, or appeal to the government or court." The result is coded as 1 for "Yes" and 2 for "No."

Taking action to change or not is the most basic different feature to measure environmental deprivation. Table 6.6 shows the distribution of taking action to change or not for all the three groups of residents. The table shows the fraction of the residents took an action is decreased from the rural residents, migrants to urban residents. This seems weird since the urban (rural) residents who have the highest (lowest) environmental consciousness among the three groups of residents have the smallest (largest) fraction of taking an action, respectively.

Table 6.6 Distribution of whether strive to change and why do not take any action for the three groups of residents.

Whether strive to change	**Rural (%)**	**Migrants (%)**	**Urban (%)**
1. Took actions to change this situation	40.0	32.0	25.4
2. Did not take any action to change this situation	60.0	68.0	74.6
Why do not take any action?			
1. Never thought about it, I can pass it	40.0	37.1	33.7
2. Thought about it but did not know what to do	25.0	31.1	37.4
3. Thought about it but too expensive or time costly	15.0	12.9	14.0
4. Thought about it but did not expect it works	20.0	19.7	15.0

To better understand this situation, we asked an additional question "why you did not take any action?" for the responders who answered, "did not take any action." The responders can choose one or more than one of the following four choices: "Never thought about it, I can pass it," "Thought about it but did not know what to do," "Thought about it but too expensive or time costly," and "Thought about it but did not expect it works." Table 6.6 shows less fraction of urban residents choose "never thought about it, I can pass it" or "thought about it but did not expect it works" and less fraction of rural residents choose "thought about it but did not know what to do." It shows the main reason that less urban residents take action is they lack knowledge of how to take action. Therefore, the publicity of the way to take action may be useful to increase the action taking rate.

In addition, is the environmental consciousness really having a negative effect on the action taking as shown in Table 6.6? To answer this question, we analyze the NEP and take action to change or not of each group of residents. We used a Chi-squared test since the independent variable NEP is a nominal variable and the dependent variable is a categorical variable. We expect to see the people who hold high environmental conscious would strive to change the surrounding environment by interacting with others.

Three groups of residents were separated for the comparison. Table 6.7 shows there is a statistically significant relationship between NEP and action or not. This result shows extremely significant for the migration group (Sig. $<.001$) and also significant for the other two groups of residents (Sig. $<.05$). Here, although the significance is less than .05

Table 6.7 Chi-squared test results of the NEP and the strive to change.

Three groups of residents			*B*	S.E.	*df*	Sig.	Exp (*B*)
Rural	Step 1[a]	NEP	− .107	.042	1	.011	.898
		Constant	5.958	2.215	1	.007	386.9
Migrant	Step 1[a]	NEP	− .084	.023	1	.000	.920
		Constant	5.406	1.355	1	.000	222.8
Urban	Step 1[a]	NEP	− .076	.031	1	.014	.927
		Constant	5.515	1.848	1	.003	248.3

[a]Variable(s) entered on step 1: NEP.

but greater than .001 for the other two groups, the differences are also interesting and require a deeper analysis. The negative *B* values for all the three groups of residents show that the analysis results support the hypothesis, that is, people with higher the level of environmental consciousness are more likely to take actions for all the three groups of residents.

6.4.2 Action approaches

For the residents who decided to take action, they may take different types of actions. What are the approaches they adopted? Can we claim the adopted approaches are following some trends? Are there any differences that would appear for the three groups of people? We will answer these questions by analyzing specific actions that people strive to change their surroundings.

The strive to change actions includes the following six specific actions, which is listed as follows: (1) Discuss with other people in surroundings; (2) Approach to neighborhood committee and/or village committee cadres; (3) Approach directly to the polluter; (4) Approach to a government agency; (5) Report the issue to mass media; and (6) Fill a lawsuit to court. In our questionnaire, we allowed the responders to choose one or more than one approach if they choose "yes" to the question "Did you take any action to change this situation? Such as caution the polluters, go to the media, or appeal to the government or court."

Table 6.8 shows the fraction of each specific action that people choose to fight for the better surrounding environment if they took any action. In the table, the difference between the rural residents, migrants, and urban residents are shown. For the rural residents, the fraction of each action decreases in the order from Approach 1 to 6. Table 6.8 shows almost half responders who took an action would discuss the issue with other people in surroundings which is the easiest action based on time

Table 6.8 Distribution of the six specific strive to change actions of the three groups of residents.

Six specific actions	**Rural (%)**	**Migrants (%)**	**Urban (%)**
1. Discuss the issue with other people in surroundings	49.9	44.6	46.5
2. Approach to neighborhood committee and/or village committee cadres	27.2	2.5	13.2
3. Approach directly to the polluters	27.0	19.1	7.2
4. Approach to government agency	18.0	10.1	18.4
5. Report issues to the mass media	4.5	18.5	21.6
6. Fill a lawsuit to the court	0.0	5.2	7.0
Total	126.6	100.0	113.9

and economic consuming. The approaches with the next two largest fraction are the approach to village committee cadres in rural areas and approach directly to the polluters. It may be because it is more efficient to solve the environmental problems and/or achieve the demands of these two approaches. It also shows that rural residents tend to solve issues from local and direct ways. Indirect approaches such as to government agencies or the mass media are less chosen. It is also worth noting that in this survey, there is no approach to the court for the rural residents. It may be because traditional Chinese pursue no suit. Master Confucius said, "In hearing litigations, I am not different than any others. What is necessary, however, is to cause the people to have no litigations." So traditional Chinese consider a lawsuit to the court as a disgrace. Overall, the rural residents are more likely to take an action locally or directly instead of to the high-level government department or broadcast it.

For the urban residents, as shown in Table 6.8, the fraction of each action decreases in the order of Approaches 1, 5, 4, 2, 3, and 6. Similar to the rural residents, there are almost half the responders who took an action would discuss the issue with other people in the surroundings. However, the following approaches are in different orders. The approaches with the second and third largest fraction are Approach 5 (report issues to the mass media) and Approach 4 (approach to a government agency), respectively. This change can be understood as follows: urban city development not only changed people's lifestyles but also shifted their way of solving problems. People in urban areas tend to avoid direct conflict like arguing with polluters in person. Although the approach to the neighborhood committee is more convenient than to the government agency or to the mass media, the neighborhood committee in an urban area does not have much power to solve problems, but mass media and government approaches are usually more efficient and could make pressure on polluters. Therefore, they inclined to reach mass media or even fill a lawsuit to the court. This also gives us a sign that mass media is playing an important role in the urban region. Mass media plays a role in contributing to environmental consciousness that cannot be ignored. We will have a chapter (Chapter 8: Environmental Consciousness Under the Tide of the Times) to discuss the impactions of mass media. Approach 6, that is, filing a lawsuit to the court also has the least fraction for urban residents. This may be due to it more time and money for people who choose this approach. Compared to rural residents, urban residents tend to use more effective approaches rather than local or direct approaches.

For the migrants who migrated from rural to urban areas, the fraction of each approach is not all in between rural and urban residents. The fraction of Approaches 2 and 4 are less than both rural and urban residents, as shown in Table 6.8. The second approach, that is, approach to neighborhood committee, even falls into the least chosen one. The reason is understandable—migrants generally are new in their neighborhood and they usually have a weak relationship with their neighborhood committee. The other approaches are similar to rural and urban residents, with the most fraction is discussing the issue with other people in surroundings and unlikely to file a lawsuit to the court. As a result, the fraction of each action decreases in the order of Approaches 1, 3, 5, 4, 6, and 2. In addition, migrants use the least approaches in total and no migrant uses more than one approach. This may also reveal that the migrants, as a special group of residents, their behavior is not a combined one of the rural and urban residents', however, has their own character, for example, they are less confident to fight for their own right compared with the other two groups of residents.

Besides the qualitative analysis of the environment deprivation, quantitatively analysis is also performed by using the total number of approaches responders chosen since the responders can choose one or more than one approach in the questionnaire. Here, we consider that Approach 1 is not a direct way of trying to solve the environmental problems, we only count the other five approaches. Each of the approaches from Approaches 2 to 5 will be counted as 1, and the maximum possible value is 5 if the responder chooses all Approaches 2–5. Regressions are then used to find the relationship between the environmental consciousness and the count of these FSAs that selected by each person. The three resident groups are

Table 6.9 Environmental consciousness impacts on the count of Five Strive Approaches (FSAs).

NEP and action counts	Rural	Migrant	Urban
	B (beta)	*B* (beta)	*B* (beta)
NEP	.043 (.348)	.013 (.009)	− .036 (−.016)
Constant	− 1.155***	.321***	2.941**
F (df)	6.054 (1)*	1.942 (1)	.32 (1)*
Adjusted R^2	.101	.007	.059

Note: ***: $P<.01$; **: $P<.05$; *: $P<.1$.

controlled to show the relations within each resident group. In the analysis, we use the FSAs (count of these Five Strive Approaches) as the dependent variable and the NEP as the independent variable. Since both these two variables are interval variables, a linear regression analysis is used for each group of residents, so the three groups of residents are controlled. The regression result shows the count of FSAs is not significantly dependent on the NEP for all the three groups of residents, as shown in Table 6.9.

6.5 Theory models for environmental behaviors

In the last section, we have discussed the relationship between environmental consciousness and environmental behaviors. However, the environmental behaviors are results of many factors, as the complexity of the underlying facts was also discovered by many previous researchers. Recently developed theories and models give hope for building an understanding of people's behaviors that could contribute to environmental issues. Meanwhile, some of them offered effective models that helped to improve our insights into the environmental behaviors in China. Three theoretical models from western society, concentrate on getting people to change their ingrained individual daily actions, are borrowed here for reviewing and further discussions.

The first theoretic model is the theory of planned behavior (TPB). It states that the person's intention to perform the environmental behavior is driven by social pressure, subjective norms, and perceived behavioral control over the action. These three kinds of consideration, that is, "behavioral beliefs," "normative beliefs," and "control beliefs" are the factors

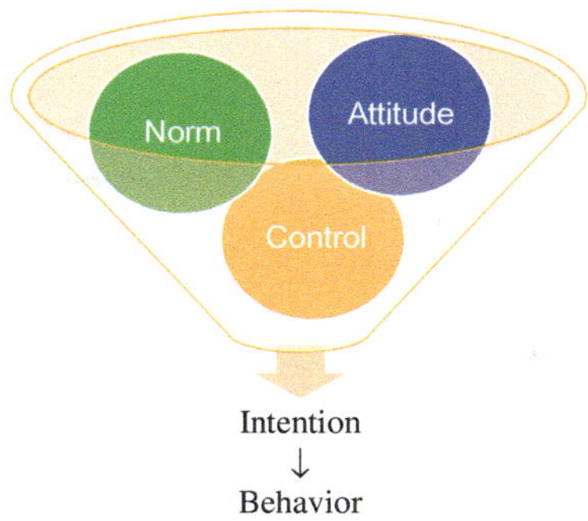

Figure 6.1 Schematic of the theory of planned behavior (TPB).

likely leading to the formation of a "behavioral intention" (Ajzen, 1985; Ajzen, 2002). A schematic illustration of the theory is shown in Fig. 6.1.

The second theoretic model is the Attitude–Behavior–Context (ABC) theory. This theory believes the Behavior (B) is a function of the Attitude (A) and the Context or external conditions (C). The strongest attitude–behavior relationship shows up when contextual factors are neutral. When contextual forces are strongly positive or negative, either all the residents are recycling or not recycling, respectively. The attitude–behavior relationship thus approaches zero (Guagnano, Stern, & Dietz, 1995; Stern, 2000).

The third theory, that is, the Value belief norm (VBN) theory, is a more advanced theory model compared with the TPB and the ABC theory. It combines the value theory, norm-activation theory, and new environmental paradigm (NEP). It contributes to building reliable variables for the contextual between attitude and behavior (Stern, 2000). The VBN theory considers the values including three dimensions: biospheric, altruistic, and egoistic. Different activities of human beings are based on these dimensions of values. The VBN theory further links the values to the belief and norm considering six variables through a causal chain: values, ecological worldview, adverse consequences, the ascription of responsibility, pro-environmental person norm, and environmental behaviors, as shown in Fig. 6.2. The causal chain begins with relatively stable general values, to a more prominent belief in the relationship between humans and the environment (ecological worldview), and then an individual's belief in the threat to the value object and the sense of obligation to take pro-environmental actions, and finally activates individuals to take positive environmental actions. (Stern, Dietz, Abel, Guagnano, & Kalof, 1999).

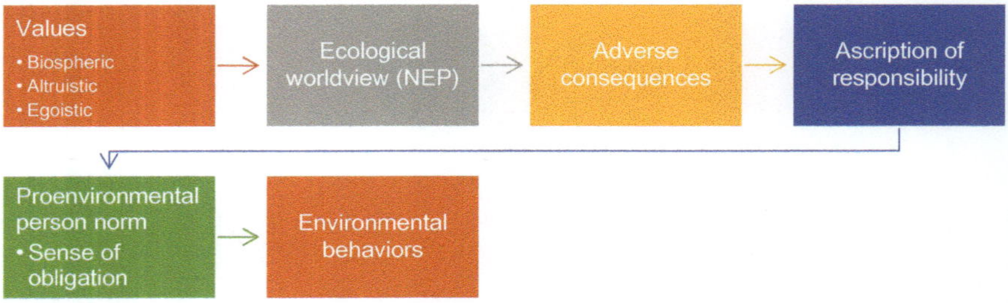

Figure 6.2 A schematic representation of variables in the VBN theory of environmentalism (Stern, 2000).

Compared with these three environmental behavioral theories, there are two commonalities: for the attitude or the values, both the TPB and the ABC theory emphasize the influence of individual attitude variables on the environmental behavior, which is expressed as values in the value–belief–normative model. Although the expressions of "values" are different from "attitudes," their connotations are positive or negative evaluations of environmental behaviors. The second commonality is external norms or conditions. Both the TPB and the ABC theory emphasize the influence of external variables on environmental behavior. The former uses external norms, while the latter uses external conditions. It should be said that the external conditions in the ABC theory visualize the external norms of the TPB. In the VBN theory, the external conditions are subdivided into adverse consequences and ascription of responsibility and the external norms are limited to pro-environmental personal norms. It should be said that the breakdown of the VBN theory is more conducive to the study and discussion of the factors influencing environmental behavior.

Considering the commonalities among those three theories and the advantages of the VBN theory, we consider VBN as our research model for the consciousness–behavior relationship explore. The VBN model gives us an opportunity to answer the question we had, that is, "do daily actions related to environmental consciousness?" It may also give the answer to more questions, such as "do beliefs and pro-environmental personal norms related to behaviors, as VBN suggested?" In our research, we simplified four key variables logically nexuses in the VBN model between values and behaviors to three, that is, ecological worldview, adverse consequences for valued objects, and ascription of responsibility were measured and analyzed.

6.6 VBN model adoption and verification

In the following sections, we will analyze the relationship between these factors to verify the influencing factors based on the VBN theory. In order to analyze these factors, we are trying to quantify the three key variables, that is, ecological worldview, adverse consequences for valued objects, and ascription of responsibility as follows:

The attitudes and values of environmental behavior can be reflected by the environmental worldview or consciousness in the VBN model, and we have discussed using the NEP to measure the environmental consciousness in Chapter 3, From Movements to Measurement.

Pollution is a negative environmental issue when people consider the consequences of an adverse impact. These considerations will further affect people's behavior. The consideration of the consequences thus reflects previous pollution experience and on the other hand, the level of the environmental consciousness. Therefore, the consciousness of consequences can be stimulated by environmental consciousness and is an indispensable part between environmental consciousness and environmental behavior. The measurement of the adverse consequences is using the answer to the question "Have you encountered environmental adverse consequences (e.g., pollution)?" Likert 5 points scale was used for the questionnaire and a higher score means more agreement with the statement.

People usually attribute the responsibility to either the individual, the manufacturers, or the government. This may lead to different environmental behaviors. Individual responsibility may directly relate to the daily actions of individuals. The manufacturer's or government's responsibility may relate to strive to change. To measure the ascription of responsibility, the question "who should take responsibility for the pollution?" was used in our research. Respondents would select one of the following options: individuals, manufactures, or the government. The selections are then recoded into two scales, that is, 1 for individual responsibility and 0 for nonindividual responsibility. The manufacturer's and the government options are both recoded as nonindividual responsibility.

After the three key factors are quantified, we then performed linear regressions between these three key factors and different types of environmental behavior. To verify the VBN theory, we consider all the environmental behavior not related to the environmental consciousness and check whether they are related to the adverse consequences and ascription of responsibility linked in between. This is done by comparing three models, that is, Model 1 only uses NEP as the independent variable, Model 2 adds adverse consequences to the independent variables, and Model 3 adds both adverse consequences and ascription of responsibility to the independent variables. If from Model 1 to Model 3, more correlations between the independent variables and the environmental behavior are activated and the adjusted coefficient of determination R^2 significantly increases to a large value, the environmental consciousness is verified to link to the environmental behavior by the adverse consequences and ascription of responsibility. In this case, the VBN theory is thus verified. Here, the adjusted coefficient of determination R^2 is used to measure how well the data is replicated by the three models. A larger R^2 means a better replication of the model.

In the following subsections, we use this method to check all the environmental behavior not directly related to environmental consciousness. As discussed in the last chapter, the WTP for urban residents and migrates, daily activities, and the FSAs meet these criteria. Therefore, they will be analyzed.

6.6.1 Willingness to pay

The comparison of the three models for WTP of urban residents and migrants are listed in Table 6.10. The table shows that the connection made by the VBN theory was not supported between WTP and environmental consciousness for both urban residents and migrants. This is because by adding "adverse consequences" and "ascription of responsibility" as the independent variables to the model one-by-one, there is still no correlations between the WTP and the independent variables for both the two resident groups. In addition, the adjusted R^2 is very small and there is no significant increase for both the two resident groups. This result is consistent with the discussion in the last chapter, that is, the residents are willing to pay when it benefits their health or finance in long terms but not related to the environmental consciousness even if adverse consequences and ascription of responsibility is stimulating.

6.6.2 Daily activities

The comparison of the three models based on the daily activities of all the three groups of residents is listed in Table 6.11. For the rural resident

Table 6.10 The comparison of the three models based on the WTP of urban residents and migrant.

Willingness-to-pay		**Model 1**	**Model 2**	**Model 3**
		***B* (beta)**	***B* (beta)**	***B* (beta)**
Migrants	NEP	.005 (.044)	.004 (.037)	.004 (.037)
	Adverse consequences	–	− .001 (−.001)	− .008 (−.009)
	Ascription of responsibility	–	–	.045 (.024)
	Constant	1.689***	1.726***	1.741***
	F (df)	.729 (1)	.245 (2)	.225 (3)
	Adjusted R^2	− .001	− .004	-.006
Urban	NEP	− .005 (−.047)	− .005 (-.050)	-.005 (-.046)
	Adverse consequences	–	− .016 (-.017)	.024 (.025)
	Ascription of responsibility	–	–	− .181 (−.105)
	Constant	2.212***	2.295***	2.156***
	F (df)	.568 (1)	.350 (2)	1.006 (3)
	Adjusted R^2	− .002	− .005	.000

Note: ***: $P<.01$; **: $P<.05$; *: $P<.1$.

group, the connection made by the VBN theory was found between the daily activities and the environmental consciousness. Model 1 was present the model without "adverse consequences" or "ascription of responsibility." There is no significant relationship between environmental consciousness and daily activities. With adding the "adverse consequences," Model 2 shows some significance of the relationship. After adding both "adverse consequences" and "ascription of responsibility," Model 3 presents a significant relationship between environmental consciousness and daily activities. The adjusted R^2 also shows the explanation level raised from 1.0% for Model 1 to 13.7% for Model 3.

For the migrant group, the connection made by the VBN theory was not found between the daily activities and the environmental consciousness. Models with or without the two independent variables, that is, "adverse consequences" and "ascription of responsibility" were not showing any significant relationship. The adjusted R^2 also did not show any significant increase.

For the urban resident group, the connection made by the VBN theory was found between the daily activities and the environmental consciousness. Similar to the rural resident group, the urban resident group

Table 6.11 The comparison of the three models based on the daily activities of all the three groups of residents.

Daily activities		**Model 1**	**Model 2**	**Model 3**
		***B* (beta)**	***B* (beta)**	***B* (beta)**
Rural	NEP	.149 (.127)	.319 (.259)**	.253 (.210)
	Adverse consequences	–	2.950 (.312)	2.034 (.219)
	Ascription of responsibility	–	–	4.865 (.275)***
	Constant	39.423***	39.703***	36.338***
	F (df)	2.747 (1)*	9.091 (2)***	6.002 (3)***
	Adjusted R^2	.010	.111	.137
Migrants	NEP	.019 (.045)	.019 (.049)	.019 (.049)
	Adverse consequences	–	−.092 (−.026)	−.110 (−.031)
	Ascription of responsibility	–	–	.203 (.030)
	Constant	34.457***	34.710***	34.672***
	F (df)	.781 (1)	.573 (2)	.341 (3)
	Adjusted R^2	.000	.002	. 007
Urban	NEP	−.023 (−.068)	−.022 (−.067)	−.018 (−.056)
	Adverse consequences	–	.401 (.121)**	.492 (.149)**
	Ascription of responsibility	–	–	−.908 (−.161)**
	Constant	34.888***	33.471***	33.404***
	F (df)	1.197 (1)	2.520 (2)*	2.552 (3)**
	Adjusted R^2	.010	.121	.204

Note: ***: $P<.01$; **: $P<.05$; *: $P<.1$.

also changes from no significance of relationships to more and more significant of relationships when adding "adverse consequences" and "ascription of responsibility" one-by-one. The explanation level also raised from 1.0% for Model 1 to 20.4% for Model 3 indicated by the adjusted R^2.

6.6.3 Strive to change the environment

The direct relationship between environmental consciousness and strive to change the environment was tested in the last chapter. The results show significant relations for action or not, however, no relation shows for the count of FSAs that people selected. Therefore, we only check whether the count of FSAs follows the VBN theory in this subsection. Similar to the last chapter, FSAs neglect the action "discuss environmental problems with others" and only count the other five actions, that is, "approach to neighborhood committee and/or village committee cadres," "approach directly to the polluters," "approach to government agency," "report issues to the mass media," and "fill a lawsuit to the court." Linear regression is then used for the three models to verify the VBN theory.

Table 6.12 shows the regression result for all three groups of residents. For the rural group, the connection made by the VBN theory was found between the count of FSAs and environmental consciousness. The Model 1 without the independent variables "adverse consequences" and "ascription of responsibility" shows no relationship between the environmental consciousness and the environmental dispute and movement. After adding the "adverse consequences," Model 2 shows some significance of the relationship between the environmental consciousness and environmental dispute and movement. By adding both "adverse consequences" and "ascription of responsibility," Model 3 presents a significant relationship between the environmental dispute and movement and both the environmental consciousness and the ascription of responsibility. From Model 1 to Model 3, an increase of the explanation level from 10.1% to 22.1% is also shown in Table 6.12.

For the migrant group, the connection made by the VBN theory was not found between the environmental consciousness and FSAs. There are no significant relation shows for all the three models. There is also no increase in the adjusted R^2 from Model 1 to Model 3. Interestingly, this result is the same as the daily activities dimension.

For the urban group, the connection made by the VBN theory was found between the environmental consciousness and FSAs. From no significant relation shows in Model 1, after adding "adverse consequences" or both "adverse consequences" and "ascription of responsibility," both Models 2 and 3 showed significant relations between the environmental consciousness and environmental dispute and movement. The explanation level is also some increasing. Therefore, for the urban resident group, the VBN theory has also received data support.

In summary, the connection made by the VBN theory was found between the environmental consciousness and both the daily pro-environmental activities and the FSAs for rural and urban groups of residents. Migrants, as a special resident group, does not show behaviors alike rural/urban groups. With this question mark, we will try to analyze this unique group in the following section.

Table 6.12 The comparison of the three models based on FSAs of all the three groups of residents.

FSAs		Model 1	Model 2	Model 3
		B (beta)	*B* (beta)	*B* (beta)
Rural	NEP	.043 (.348)	.053 (.019)**	.065 (.019)***
	Adverse consequences	–	− .025 (.147)	− .071 (.142)
	Ascription of responsibility	–	–	− .854 (.404)**
	Constant	− 1.155***	− 1.513***	− 1.806***
	F (df)	6.054 (1)*	3.962 (2)**	4.401 (3)***
	Adjusted R^2	.101	.141	.221
Migrants	NEP	.013 (.009)	.014 (.009)	.014 (.009)
	Adverse consequences	–	− .065 (.090)	− .061 (.094)
	Ascription of responsibility	–	–	− .024 (.174)
	Constant	.321***	4.98***	.493***
	F (df)	1.942 (1)	1.437 (2)	.957 (3)
	Adjusted R^2	.007	.007	.001
Urban	NEP	− .036 (−.016)	− .030 (.017)*	− .031 (.016)*
	Adverse consequences	–	− .309 (.193)	− .252 (.196)
	Ascription of responsibility	–	–	− .352 (.267)
	Constant	2.941**	3.841 ***	3.751 ***
	F (df)	0.32 (1) *	3.757 (2) **	3.116 (3) **
	Adjusted R^2	.059	.083	.096

Note: ***: $P<.01$; **: $P<.05$; *: $P<.1$.

6.7 Conclusion

The research on environmental consciousness is to better study the environmental behaviors. The investigation and evaluation of the residents' environmental consciousness are not only to understand the environmental consciousness and the environmental understanding of residents, but also to propose countermeasures to improving the environment with a scientific basis. The environmental behavior theory considers the impacts of environmental consciousness. However, much previous research found that the relationship between environmental behavior and environmental consciousness is not as close as people thought. They cannot found a persistent significance between them. Some researchers even argued: the relationship between environmental behavior and environmental consciousness is uncertain (Sun & Wu, 2007).

It is generally believed that people generally start from a lower degree of environmental behavior and gradually improve the level of environmental behavior. This is a gradual and continuous improvement process. In our research, environmental behavior was divided into three levels. The first level is the willingness, includes willing to change behaviors to pro-environmental activities and willing to pay behaviors for the pro-environmental products. According to Stern's definition, willingness was to present the essence of "intention." Our research results of willingness-to-change of all the three groups of residents and WTP of rural residents support the hypothesis, that is, people who hold high environmental consciousness would have a high willingness.

Stern's definition of environmental behavior has another innate character, that is, "impact." As the findings in previous research approve the differences between the environmental behaviors at the personal level and the public level, two more levels were divided into our research. They are, the second level, personal environmental behavior in daily activities, which covers the activities on the personal level; and the third level, strive to change the environments, which covers the activities at the public level. The uncertainty of the relationship was repeated when we conducted our analysis of our data with these two levels. No direct relationship is shown within daily activities data, and the strive to change the environments also get challenged.

After further analysis, we believe that the reason for the significant relationship between the willingness-to-change and the environmental consciousness may be because they both are abstract parts. For the values part of the environmental consciousness, it has a significant influence on the willingness-to-change. However, in the environmental activities and strive to change the environments, the questionnaire asks the actual activities. Although the actual activities and the values are closely related, actual activities are influenced by the situations and have different reactions for different levels of the actual activities. For example, in the strive to change the environments, the majority of the respondents choose "discuss with other people surroundings." Much fewer respondents choose "approach directly to the polluter," "approach to government agency," and/or "report issue to mass media." This may cause a direct connection between the environmental activities and the environmental consciousness not significant.

This not significant relationship may as the previous researchers said, "the socio-economic and environmental consciousness of China has not yet reached the level to affect the people's pro-environmental behavior." Although this cognition has some truth, it is superficial. In theory, consciousness is the foundation of the behavior, however, environmental protection must be coordinated with social, economic, and other conditions to the realize from consciousness to activities.

After the direct impact of the environmental consciousness on environmental behavior is diluted by uncertainties of the situations, we want to know what factors affect in the process between the environmental consciousness and environmental behavior. The VBN theory model was thus introduced for further analysis. Following two factors, that is, beliefs and pro-environmental personal norms, from Stern's model, three factors, that is, the ecological worldview or the NEP, the adverse consequences for valued objects, and the ascription of responsibility to take pro-environmental actions were measured and analyzed in our research.

Generally, the VBN model has been approved in our research. As our research separated people into three groups: the rural resident group, the urban resident group, and the migrant group, the analysis was done within each of those three groups. The results show that the rural and urban resident groups supported the VBN model. For the regression of only the NEP and the environmental behavior with no significant relationship, after adding "adverse consequences" and "ascription of responsibility," the environmental consciousness is stimulated. The relationship between the NEP and the environmental behavior becomes significant and the explanatory level has also increased.

The migrant group of residents still has an unproven relationship between their environmental behavior and environmental consciousness. The data of environmental activities cannot support the VBN theory, however, based on the particularity of the migrants and our investigation, the ambiguity of "consciousness" and "responsibility attribution" also implies the verification of the VBN theory. As a result, the environmental conditions of environmental behavior are also an important part of environmental protection work. As more and more migrants in China due to the economic transformation discussed in Chapter 4, From a Longitudinal Study to Cross-Sectional Study, they cannot be overlooked in the coming era, and their environmental behavior could be the main power of the CHANGE.

CHAPTER SEVEN

The path to environmental behaviors

7.1 Theory reviews of the demographic variables

Before we construct the model for the environmental consciousness, we need to carefully choose the demographic variables to start with. Here, we review the general conclusions from the previous research which mainly considered the relations between the environmental consciousness and the demographic variables such as sex, age, education, and place of residence.

7.1.1 Sex

The early research paid attention to the sex differences in environmental consciousness from the 1970s but did not reach a consensus conclusion. McEvoy believes that men are more concerned about the environment because men are more likely to be politically active, more involved in community issues, and have a higher level of education than women (McEvoy, 1972). Passino and Lounsbury believe that men are more concerned about their work and financial savings than women, so they are less concerned about protecting the environment than women (Passino & Lounsbury, 1976; van Liere & Dunlap, 1980). Since the mid-1980s, research on the relations between sex and environmental consciousness has increased, and gender differences in environmental consciousness have become an interesting topic of more and more research (Davidson & Freudenburg, 1996). In the early 1990s, Mohai pointed out that the relationship between sex and environmental awareness still uncertain. Some studies have found that women are more concerned about the environment than men and have higher environmentally consciousness. They are more likely to develop a sense of crisis and insecurity in environmental problems (Stern et al., 1993). Others have found that men are more concerned about the environment than women (Davidson & Freudenburg, 1996). There are also studies showing no sex differences in environmental

Environmental Consciousness in China.
DOI: https://doi.org/10.1016/B978-0-08-100388-6.00007-6

consciousness (Mohai, 1992). In recent studies, women are concluded to have higher environmental consciousness than men. It is also reflected in women's friendly behavior in the environment. So, some researchers believe that women's more concern about the environment than men has become a conclusion of the increasingly widespread environmental sociology (Tindall, Davies, & Mauboules, 2003).

The same structural perspective is used to study the sex differences in environmental consciousness in China and Hong pointed out that the tradition of "men should work outside while women should stay at home" in Chinese culture does not lead women to pay attention to environmental protection. Men are expected to and believe that they should pay attention to environmental protection. Moreover, the separation of the roles of men and women in Chinese society is limited. Chinese society has been a collective standard rather than an individual-based society for a long time. Especially for women, it is difficult for them to express their opinions independently. Their opinions are often influenced by collective interests. Chinese women are also more concerned about employment and financial savings than environmental protection considering the family's interests. In addition, Chinese society has not experienced the impact of feminist movements in Western society, and the mobilization of women is very limited (Hong & Xiao, 2007). A survey conducted by Shen and Saijo in China shows that men are more concerned about the environment than women (Shen & Saijo, 2008).

7.1.2 Maternal character and parenthood

Based on sex, scholars have also proposed "maternal character." Sociological theories believe that individual attitudes and behaviors can be predicted through the process of socialization, which shapes the different social roles of men and women. In many cultures, the socialization of women makes them develop the role of giving love. In other words, it is the role of having and taking care of children. Therefore, the process of women's socialization encourages them to be more compassionate and more capable of fostering, protecting, and cooperating. Once these concepts are internalized, they form the "maternal character" of women. It shows women have stronger ethics of care and expand it into a protective attitude toward nature. This is because women see themselves as a part of the community and the bigger society. In realistic analysis, the "maternal character" hypothesis may be verified by that the woman who has 0- to

6-year-old children at home may have higher environmental consciousness than other residents, especially men. In addition, since there consider having children or not, this assumption is also the assumption of the partiality to their children. Hamilton made it clear that age, sex, and parenthood are three important demographic variables that predict environmental consciousness, and women with children concern the environmental protection the most (Davidson & Freudenburg, 1996).

7.1.3 Age

Almost all studies have found that age is closely related to environmental consciousness. Van Liere and Dunlap found that young people have higher environmental consciousness than older people. This may be because the ecological values are contrary to traditional concepts. Traditional ideas such as utilitarianism or anthropocentrism have not yet become the mainstream in the young people's concepts since they have not been fully socialized. So, young people are more susceptible to new worldviews and values such as environmentalism. Another reason may be that young people are more likely to be attracted to information related to environmental issues than older people (van Liere & Dunlap, 1980). Other researchers believe that with the development of socialization, the environmental consciousness of the elder people is actually strengthening. This may be due to the increasing involvement of the mass media in environmental issues (Eagly & Kulesa, 1997). Some researchers have concluded that elderly people are more concerned about environmental issues in China. Shen and Saijo believe that older people are more concerned about environmental issues and are more tend to have a positive environmental attitude after surveying in Shanghai (Shen & Saijo, 2008). This may be due to the fact that the elderly people experienced serious environmental pollutions from the early 1980s to the early 1990s. Another possible explanation is that Chinese parents have cared about their children over themselves since ancient times. The one-child policy has strengthened this concept. Therefore, the concept of maintaining and creating a good environment for future generations makes older people have higher environmental consciousness than younger people.

7.1.4 Education

The relationship between the education level and environmental consciousness is quite important in demographic variables. The impact of the education level on environmental consciousness has been confirmed in almost all relevant research. Researchers have shown that residents with a higher level of education tend to have higher environmental consciousness levels. In other words, the education level and environmental consciousness are positively correlated (Scott & Willits, 1994). Surveys in China also confirm this conclusion is valid in China (Shen & Saijo, 2008). For this general conclusion, the researchers have the following explanation: the education includes knowledge related to environment and environmental protection. At the same time, education may also improve the understanding of environmental issues.

7.1.5 Place of residence

The relationship between the place of residence and the environmental consciousness is the main focus of this book. Previous research has compared the environmental consciousness of urban and rural residents. The study found that urban residents have a higher level of environmental consciousness. Fransson and Gärling believe one possible reason may be that urban residents are more exposed to harsh environments, such as air pollution. These direct experiences of the profit and loss analysis of their interest will enhance their environmental consciousness (Fransson & Gärling, 1999). This result is similar to Hong and Lu's recent research in China based on the Chinese General Social Survey in 2003 (CGSS2003). Hong and Lu found: "In the past several decades, many researchers focused on individual-level determinants of environmental concern while neglected the effects of regional-level factors on environmental concern. The findings indicate that primary industry output to GDP ratio and industrial soot emissions are significantly related to levels of environmental concern. The higher the primary industry output to GDP ratio, the lower is the level of environmental concern; the more industrial soot emissions produced by the city, the higher is the level of environmental concern" (Hong & Lu, 2011).

7.1.6 Other demographic variables

Other than the demographic variables discussed above which are considered related to the environmental consciousness, some researchers have also found that political parties and religious beliefs are also closely related to environmental consciousness. In this research, we slightly modified the selection of the demographic variables based on the previous research on the relationship between the demographic variables and the environmental consciousness and the national conditions of China as follows.

Whether the only child is added since during the survey, we noticed that the age of the migrants is basically between 16 and 30 years old. The residents in this age group are the generation that the one-child policy starts. We also considered the difference in the execution of this policy in urban and rural areas, so "whether is the only child" variable is added. We believe that the only child in the family is more likely to be treated as the core of the family, and is thus more inclined to personal centralism or anthropocentrism in the concept of the environmental consciousness, and may thus have a lower environmental consciousness level.

Whether a chronic disease is also added since the health condition may also have an impact on attitudes. This is inspired by the Quality of Life (QOL) scale. We believe that the environmental blame of patients with chronic diseases may enhance their environmental consciousness. Therefore, we include this variable in our demographic variables.

In addition, based on the national conditions in China, we did not include the political party variable due to the single-party system in China.

Lastly, the place of residence variable is further combined with two more variables, that is, age and the residence time at the place and the migration experience is thus obtained. Based on the migration experience, the respondents are divided into the urban resident group, the rural resident group, and the rural-to-urban migrants group, as discussed in Chapter 4: From a Longitudinal Study to Cross-Sectional Study.

In summary, there are eight variables that are included in our demographic variables, that is, sex, age, education level, religious beliefs, residential groups, whether having 0- to 6-year-old children, whether the only child, and whether a chronic disease. Within these eight demographic variables, we will split the analysis among the three residential groups and perform regression on the rest seven demographic variables.

7.2 Model construction of the environmental consciousness

Chapter 5: What Drives Environmental Consciousness? discussed the influence of the three driving models, that is, the pollution driving model, the constructivism model, and the postmaterial model on environmental consciousness. Based on the understanding of the factors affecting the environmental consciousness discussed in the previous chapters, this section will comprehensively analyze the three groups of residents and construct models for the relations.

In the analysis, seven demographic variables discussed in the last section and nine (10 for migrants) variables for the three driving models are used as the independent variables. The NEP scores are used as the dependent variable. In each of the analysis, we compare three models, that is, Model-D: the independent variables use only the demographic variables, Model-T: the independent variables use only the nine (10 for migrants) variables for the three driving models, and Model-A: the independent variables use all the independent variables used in Model-D and Model-T. The explanatory level or the adjusted R^2 is compared for the three models and based on this result, we construct new models for each group of residents.

7.2.1 Rural residents

The regression result of the three models for rural residents is shown in Table 7.1. For rural residents, the explanatory power of Model-D is already very high, which reaches 23.2%. For Model-T when changing the independent variables to the three driving models, the explanatory power is reduced. After incorporating demographic variables and three theoretical models, that is, Model-A, the explanatory power increases to 28.1%. It should be said that the overall explanation is stronger for Model-A.

However, compared to Model-D, Model-A with the three theoretical models added only have around 5% increase in explanatory power. Therefore, the explanatory power of the theoretical dimension is not high, and the sex and education level of the demographic variables have a great influence on the environmental consciousness level of the rural residents. This may generally explain the change of the environmental consciousness, that is, for the rural population, the younger residents, the higher the level of environmental consciousness compared with the elderly residents. Residents with higher levels of education also have higher levels of environmental consciousness than those with lower education levels. For demographic variables, the data supports predecessors' research that age and environmental awareness levels are negatively correlated, while education and environmental awareness levels are positively correlated.

Table 7.1 The comparison of the models of the environmental consciousness with different factors for the rural residents.

Environmental consciousness for the rural residents	**Model-D**	**Model-T**	**Model-A**
	B **(β)**	***B*** **(β)**	***B*** **(β)**
Sex	− .905 (−.060)		− .769 (−.051)
Age	− .176 (−.396)***		− .158 (−.352)***
Education	.284 (.174)*		.178 (.109)
Only child?	1.743 (.085)		1.562 (.077)
Has 0- to 6-year-old children?	1.856 (.089)		.393 (.019)
Religion	− 1.248 (−.076)		− .905 (−.055)
Has chronic disease?	− 1.497 (−.058)		− .768 (−.030)
Pollution experience		− 3.794 (−.218)***	− 2.734 (−.156)*
Physical environment		.442 (.132)	.507 (.149)*
Knowledge of the environment		.116 (.065)	.081 (.031)
Opportunities to obtain information		.362 (.101)	.060 (.017)
Faith in science and technology		.993 (.225)***	.763 (.165)**
Housing conditions		− .414 (−.129)	− .102 (−.031)
Financial conditions		.316 (.125)	.298 (.116)
Medical and social security conditions		− .431 (−.140)	− .226 (−.072)
Traffic conditions		− .421 (−.126)	− .545 (−.161)*
Constant	49.664***	44.647***	47.117***
F (df)	7.440***	3.851***	4.461***
Adjusted R^2	.232	.137	.281

Note: ***$P<.01$; **$P<.05$; *$P<.1$.

We then optimize the model for rural residents. After using seven demographic variables and nine variables under the three theoretical dimensions as the independent variables, only the age variable is significant in the demographic variables and the education level is not significant anymore. The author believes that this may due to the education level among rural residents is not well distinguished, that is, more than 90% of the rural respondents only have a high school or lower education.

For the theoretical dimensions, the influence of the traffic conditions and the physical environment becomes significant while the significance of the pollution experience and the faith in science and technology decreases when comparing Model-A to Model-T. This decrease may due to that the traffic conditions and the physical environment are related to the experience of environmental pollution. Respondents who perceive worse environmental conditions and/or traffic conditions often more sensitive to environmental pollutions. Therefore, it may be understandable that the significance of the traffic conditions and the physical environment increases while the significance of the pollution experience decreases.

Overall, Model-A with both demographic variables and the three theoretical dimensions considered shows that rural residents with younger ages, had pollution experience, better feelings on the environmental conditions, poorer traffic conditions, and believe that science and technology can improve the environment are tending to have a higher level of environmental consciousness.

In order to better reflect the relationships between the independent variables and the environmental consciousness for rural residents, we optimized the model by using stepwise regression during the linear regression, as shown in Table 7.2. The regression result shows only age, faith in science and technology, and education are highly correlated with the environmental consciousness for the rural residents.

7.2.2 Migrants

Table 7.3 shows the comparison result of the three different models for the migrants. If only consider the demographic variables, that is, Model-D, the explanatory power is relatively low, that is, 10.5%. When only consider the three theoretical dimensions, that is, Model-T, the explanatory power has reached 24.0%, which indicates the demographic variables of the migrants are less related to the environmental consciousness than the theoretical dimensions. After incorporating demographics variables

Table 7.2 The optimized model of the environmental consciousness for the rural residents.

Environmental consciousness for the rural residents	**Optimized model**
	B **(β)**
Age	− .141 (−3.654)****
Faith in science and technology	.998(2.944)***
Education	.310(2.210)**
Constant	40.450****
F (*df*)	18.187 (3)****
Adjusted R^2	.266

Note: ****$P<.0001$; ***$P<.01$; **$P<.05$; *$P<.1$.

and the three theoretical dimensions, that is, Model-A, the explanatory power further increases to 31.3%. This result shows that the demographic variables and theoretical dimensions may generally explain the environmental consciousness of the migrants.

In Model-A, the data support previous studies for demographic variables in which age (education level) and the environmental consciousness levels are negatively (positively) correlated, respectively. In addition, another hypothesis in the last section was validated for the migrant group, that is, whether is the only child has a negative effect on the environmental consciousness. It also falsifies the hypothesis and shows that residents with better health tend to have a higher environmental consciousness.

For the three theory models, the faith in science and technology of migrants is consistent with the rural residents, that is, the residents believe science and technology tend to have a higher level of environmental consciousness. Although these conclusions are different from the previous research from other countries, this is not surprising. This may because the residents who believe in science and technology often have a higher education. In addition, pollutions in China has not shackled China with the hat of "Science and technology lead to over-development of industry" for now. As a result, the residents' belief in science and technology is still rising. Another factor in the three theory models, the perceived relative pollution is a special factor for the migrant group. It may relate to the pollution experience and its main consideration is the change in the level of the objective environment before and after the migration. The larger it decreases, the residents it predicted to pay more attention to the environment and thus may have a higher level of environmental consciousness. The data shown in Table 7.3 also validated this assumption.

Table 7.3 The comparison of the models of the environmental consciousness with different factors for the migrants.

Environmental consciousness for the migrants	Model-D	Model-T	Model-A
	B (β)	*B* (β)	*B* (β)
Sex	.492 (.033)		.776 (.051)
Age	− .031 (−.059)		− .060 (−.116)**
Education	.729 (.330)***		.538 (.238)***
Only child?	1.334 (.075)		2.314 (.128)**
Has 0- to 6-year-old children?	− .061 (−.002)		− 1.180 (−.046)
Religion	− .410 (−.016)		− .246 (−.009)
Has chronic disease?	− .559 (−.028)		− 1.785 (−.090)*
Pollution experience		− 2.448 (−.162)***	− 2.176 (−.142)***
Physical environment		.089 (.030)	.213 (.071)
Knowledge of the environment		.141 (.041)	.104 (.030)
Opportunities to obtain information		− .099 (−.033)	− .125 (−.041)
Faith in science and technology		.383 (.105)**	.423 (.115)**
Housing conditions		.009 (.004)	− .039 (−.017)
Financial conditions		.220 (.113)*	.251 (.128)**
Medical and social security conditions		− .009 (−.003)	− .015 (−.005)
Traffic conditions		− .092 (−.041)	− .205 (−.092)
Relative pollution		− .249 (−.446)***	− .232 (−.417)***
Constant	45.607***	52.304***	45.223***
F (df)	6.887***	11.381***	9.446***
Adjusted R^2	.105	.240	.313

Note: ***$P<.01$; **$P<.05$; *$P<.1$.

Overall, after controlling the seven demographic variables and ten independent variables under three theoretical dimensions, Model-A shows that migrants with younger ages, higher education level, not the only child, without chronic disease, had pollution experience, believe that science and technology can improve the environment, better financial conditions, and perceived more relative pollution are tending to have a higher level of environmental consciousness.

In order to better reflect the relationships between the independent variables and the environmental consciousness for migrants, we optimized the model by using stepwise regression during the linear regression, as shown in Table 7.4. The regression result shows only relative pollution, education, pollution experience, and faith in science and technology are highly correlated with the environmental consciousness for the migrants.

7.2.3 Urban residents

The regression result of the three models for rural residents is shown in Table 7.5. For rural residents, the explanatory power of Model-D which only considers the demographic variables is very low, that is, only 3.2%. When only consider the three theoretical models, that is, Model-T, the explanatory power increases to 12.1%. This indicates that for urban residents, the demographic variables have less impact than the three theoretical dimensions. When both the demographic variables and the three theoretical dimensions are included, the explanatory power further increases to 15.4%. Although the overall explanatory power is still not very high, it should be acceptable. Once again, the important role of the theoretical dimensions in explaining the environmental consciousness for urban residents is verified.

Since the explanatory power of the model considers the three theoretical dimensions, that is, Model-T is nearly 10% higher than the model only considers demographic variables, that is, Model-D, it is considered that the demographic variables may have much less correlation with the environmental consciousness than the three theoretical dimensions. For the demographic

Table 7.4 The optimized model of the environmental consciousness for the migrants.

Environmental consciousness for the migrants	**Optimized model** *B* (β)
Relative pollution	− .219 (− .395)****
Education	.570 (.252)****
Pollution experience	− 2.050 (− .134)***
Faith in science and technology	.405 (.110)**
Constant	46.813****
F (df)	34.032 (4)****
Adjusted R^2	.296

Note: ****$P<.0001$; ***$P<.01$; **$P<.05$; *$P<.1$.

Table 7.5 The comparison of the models of the environmental consciousness with different factors for the urban residents.

Environmental consciousness for the urban residents	Model-D	Model-T	Model-A
	B (β)	*B* (β)	*B* (β)
Sex	.037 (.002)		−.435 (−.029)
Age	−.117 (−.153)**		−.089 (−.116)*
Education	.488 (.194)***		.363 (.144)**
Only child?	1.284 (.075)		.836 (.048)
Has 0- to 6-year-old children?	.395 (.014)		−.356 (−.012)
Religion	1.592 (.033)		2.411 (.052)
Has chronic disease?	.504 (.027)		.475 (.025)
Pollution experience		−3.167 (−.211)***	−3.143 (−.209)***
Physical environment		.082 (.027)	.007 (.002)
Knowledge of the environment		1.016 (.277)***	1.208 (.325)***
Opportunities to obtain information		.035 (.011)	−.081 (−.026)
Faith in science and technology		.457 (.120)*	.442 (.115)*
Housing conditions		−.156 (−.067)	−.205 (−.089)
Financial conditions		.194 (.096)	.190 (.094)
Medical and social security conditions		−.149 (−.053)	−.241 (−.087)
Traffic conditions		−.335 (−.151)*	−.136 (−.062)
Constant	51.220***	53.096***	50.488***
F (df)	2.165***	4.664***	3.633***
Adjusted R^2	.032	.121	.154

Note: ***$P<.01$; **$P<.05$; *$P<.1$.

variables, the data support earlier research that age and environmental consciousness are negatively correlated, while education level and environmental consciousness are positively correlated. Moreover, the positive correlation between environmental knowledge and environmental consciousness is consistent with Hong and Xiao's research. In addition, the residents who have pollution experience tend to have a higher environmental consciousness. This is consistent with our hypothesis and verifies our hypothesis. However, the faith in science and technology and the environmental consciousness of the urban residents are positively correlated. This result is inconsistent with the research in other countries while the same with the other two resident groups in China. This may also be understood similar to the other two resident groups as discussed in the last subsection.

Overall, after controlling the seven demographic variables and nine independent variables under three theoretical dimensions, Model-A shows that the urban residents with younger ages, higher education level, had pollution experience, has more knowledge of the environment, and believe that science and technology can improve the environment are tending to have a higher level of environmental consciousness.

In order to better reflect the relationships between the independent variables and the environmental consciousness for the urban residents, we optimized the model by using stepwise regression during the linear regression, as shown in Table 7.6. The regression result shows only education, pollution experience, and knowledge of the environment are highly correlated with the environmental consciousness for the urban residents.

7.3 Model construction of environmental behavior

7.3.1 Rural residents

In the last section, we have found the environmental consciousness is mainly affected by age and the education level in the socio-demographic

Table 7.6 The optimized model of the environmental consciousness for the urban residents.

Environmental consciousness for the urban residents	**Optimized model**
	B **(β)**
Education	.385 (.153)***
Pollution experience	− 3.127 (−.208)****
Knowledge of the environment	1.125 (.303)****
Constant	52.291****
F (df)	11.870 (3)****
Adjusted R^2	.159

Note: ****$P<.0001$; ***$P<.01$; **$P<.05$; *$P<.1$.

variables for the rural residents. We also have found the environmental consciousness may be driven by faith in science and technology for the rural residents. In the last chapter, we also find the significant relations between the environmental consciousness and willingness, whether strive to change for the rural residents. The relationships between the daily activities, the count of FSAs, and the environmental consciousness is stimulated by adding the "adverse consequences" and "sense of obligation" based on the value-belief-norm (VBN) theory. By combining them together, we make a schematic plot of the environmental consciousness and environmental behaviors of the rural residents, as shown in Fig. 7.1.

7.3.2 Migrants

Since there is no correlation between the environmental behavior and environmental consciousness for migrants even if the VBN theory is considered, in this subsection, we attempt to construct a new model for the migrant's environmental consciousness and environmental behavior based on the analysis.

For the migrants, we believe that the time and space transition of the migrants enables them to improve their environmental consciousness fast. In the survey, we found that the environmental consciousness of migrants seems to be sensitive to the length of time after the move to the urban areas. With a longer time live in urban areas, environmental consciousness is higher. To confirm this, we look back to the analysis of the demographic variables discussed in Chapter 4: From a Longitudinal Study to Cross-Sectional Study. The age and education level plays an important role in the level of environmental consciousness. By combining these factors, we expect the longer the time they live in urban areas, the younger the age, and the higher the education level, the higher the level of

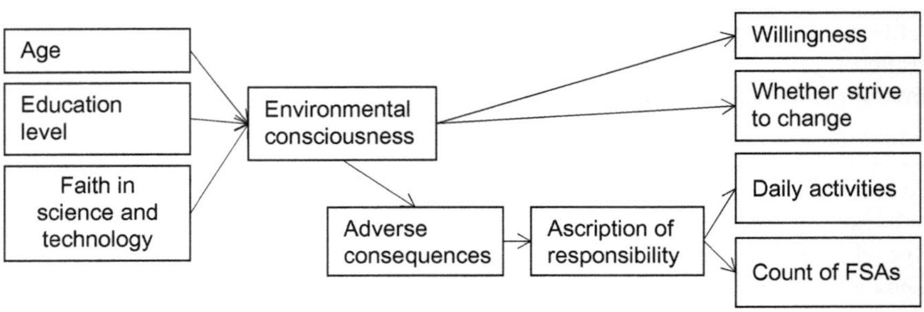

Figure 7.1 Schematic plot of the environmental consciousness and environmental behaviors of the rural residents.

environmental consciousness. Here, we use a linear regression between the urban areas living time and the environmental consciousness to verify this prediction. The age factor, which is also a time scale, is also brought into the analysis as a control variable.

Table 7.7 shows the regression analysis results with the age control. These results validated the predictions that environmental consciousness is significantly negatively correlated with age, and significantly positively correlated with the living time in urban areas. As a result, we should incorporate the living time in urban areas into the systematic model analysis for the migrants.

Although the migrants have fast increased in their environmental consciousness, the ambiguity in the "adverse consequences" and "ascription of responsibility" makes the environmental behavior not be stimulated by the environmental consciousness. During the survey, we also found that the environmental behavior of the migrants is affected by both the knowledge of the rules in urban areas and the habitually. For example, there are many migrants who are unable to know the environmental rules of the urban areas due to the lack of knowledge and language barriers. Lifestyles in urban and rural areas are very different in China. In rural areas, poor facilities cause residents living there to have their own ways to live with low cost. There are also fewer rules in public places in rural areas due to the less density of population. After the migrants move to urban areas, these new migrants start to use modern facilities in large cities. However, they may still use their rural ways to live with low cost due to the habitually. From Table 6.4, we can see the migrants still have less fraction of people re-use plastic bags, recycle waste paper, plastic bottles, batteries, and use own basket/shopping bag for daily purchase than the urban residents although a great improvement can be seen compared with the rural residents.

Table 7.7 The regression model for the living time in urban areas and the environmental consciousness for the migrants.

Urban living time and NEP	**Model**
	B **(β)**
Urban living time	.118 (.045)**
Age	− .125 (.036)***
Constant	59.561***
F (df)	5.977 (2)**

Note: ***$P<.01$; **$P<.05$; *$P<.1$.

On another aspect, the migrants may not know the consequence and responsibility in urban areas due to the lack of knowledge. Without really knowing the consequence and responsibility, some of the migrants living in urban areas for a long time just try to learn from the urban residents. Thus, the bad habits of some urban residents have a very negative impact on some of the migrants' enthusiasm for environmental behavior. These migrants may think that the residents in the urban areas even do not follow the rules, they may not follow these rules either. They are also not familiar with the urban management process, which also makes them follow the habits in rural areas. It should be said that before the "adverse consequences" and "ascription of responsibility" are not clearly established, the uncertainty of their environmental behavior may be more related to the knowledge of the rules in urban areas and the habitually instead of the environmental consciousness. This also implies the verification of the VBN theory.

Based on the analysis result of the urban living time and environmental consciousness and the aforementioned two aspects that break the relations between the environmental consciousness and environmental behavior, we included the urban residence time in the system model analysis of the migrant group, and hope it may improve the regression. In addition, according to the pollution-driven model of the migrants, we also added the pollution experience in the analysis. The unique variable of the migrants, that is, the relative pollution of the migrants is also added in the analysis to perform a deep analysis between the environmental consciousness and environmental behavior. We will perform the analysis of two types of environmental behavior, that is, daily activities and FSAs of the strive to change.

In the last chapter, we found the environmental consciousness is not related to daily activities. We also found using the VBN model for the migrant data, the environmental consciousness and the daily activities are also not related. Here, we compare four models to analyze the relations when the urban live time, the pollution experience and relative pollution is considered. The first model, Model I, directly analysis the relationship between environmental consciousness and daily activities. The next three models, Model IV, V, and VI add the urban living time, pollution experience and the relative pollution one by one. Linear regression is then used for all these four models.

The comparison of the four models based on the daily activities migrants are listed in Table 7.8. The table shows a connection was found between the daily activities and the other four variables. Model I represent the model without the urban living time, the pollution experience, and the relative pollution variables. There is no significant relationship

Table 7.8 The comparison of the four models based on the daily activities of the migrants.

Daily activities	**Model I**	**Model IV**	**Model V**	**Model VI**
	B **(β)**	***B*** **(β)**	***B*** **(β)**	***B*** **(β)**
NEP	.019 (.045)	− .019 (−.047)	− .023 (−.055)	− .051 (−.100)
Urban living time	—	.025 (.095)*	.027 (.101)*	.190 (.401)***
Pollution experience	—	—	− .129 (−.022)	− .444 (−.070)
Relative pollution	—	—	—	.014 (.067)
Constant	34.457***	34.208***	34.537***	35.734***
F (df)	.781 (1)	2.122 (2)	1.688 (3)	3.506 (4)**
Adjusted R^2	.000	.006	.006	.139

Note: ***$P<.01$; **$P<.05$; *$P<.1$.

between environmental consciousness and daily activities. By adding the urban living time, Model IV shows some significance of the relationship. After adding the urban living time and the pollution experience, the significance and the adjusted R^2 in Model V is pretty similar to Model IV. With the relative pollution added, as shown in Model VI, the significance is increased and in addition, the adjusted R^2 also shows the explanatory power raised from 0.0% for Model I to 13.9% for Model VI.

For the count of FSAs of the strive to change, similar to the daily activity, four models are compared and the urban live time, the pollution experience and relative pollution is considered. The comparison result of the four models based on the count of FSAs of the migrants is listed in Table 7.9. The table shows the connection was found for the count of FSAs. Model I was present the model without the urban living time, the pollution experience and the relative pollution variables. There is no significant relationship between environmental consciousness and the count of FSAs. With adding the urban living time, Model IV still shows no significance of the relationship. After adding the urban living time and the pollution experience, significant relation is shown between the pollution experience and the count of FSAs. The adjusted R^2 also shows the explanatory power raised from 0.7% for Model I to 14.5% for Model V. With the relative pollution added, as shown in Model VI, the significance is lost and in addition, the adjusted R^2 also shows a low explanatory power.

To conclude all the above findings, we find the relations between the environmental consciousness and the daily activities and the count of FSAs are

Table 7.9 The comparison of the four models based on the count of FSAs of the strive to change of the migrants.

FSAs	**Model I**	**Model IV**	**Model V**	**Model VI**
	***B* (β)**	***B* (β)**	***B* (β)**	***B* (β)**
NEP	.013 (.009)	−.002 (−.033)	−.001 (−.026)	.003 (.069)
Urban living time	—	.001 (.028)	.000 (.004)	−.006 (−.141)
Pollution		experience	—	—
	−.098 (−.183)			
−.291 (−.389)***				
Relative pollution	—	—	—	.001 (.055)
Constant	.321***	.259*	.720***	.169
F (df)	1.942 (1)	.346 (2)	21.902 (3)***	.641 (4)
Adjusted R^2	.007	−.003	.145	−.024

Note: ***$P<.01$; **$P<.05$; *$P<.1$.

not following the VBN theory for the migrants. However, by using a different model, the relationships between the daily activities and the environmental consciousness are stimulated by adding the urban living time, the pollution experience and the relative pollution, and the relationships between the count of FSAs and the environmental consciousness is stimulated by adding the urban living time and the pollution experience. In addition, in the last section, we have found the environmental consciousness is mainly affected by the education level in the socio-demographic variables and relative pollution, pollution experience, and faith in science and technology in the three theoretical dimensions for the migrants. In this section, we also found the environmental consciousness is related to the urban living time. In the last chapter, we also find the significant relations between the environmental consciousness and the willingness-to-change, whether strive to change for migrants. By combining them together, we make a schematic plot of the environmental consciousness and environmental behaviors of the migrants, as shown in Fig. 7.2.

7.3.3 Urban residents

For the urban resident group, we have found the environmental consciousness is mainly affected by the education level in the socio-demographic variables and pollution experience and knowledge of the environment in the three theoretical dimensions in the last section. In the last chapter, we also find the significant relations between the environmental consciousness and

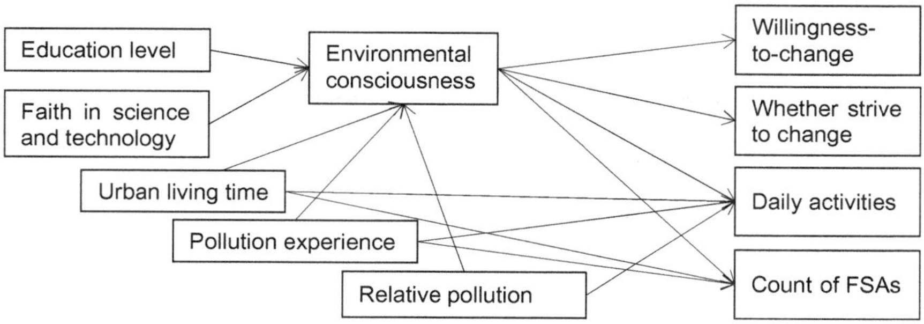

Figure 7.2 Schematic plot of the environmental consciousness and environmental behaviors of the migrants.

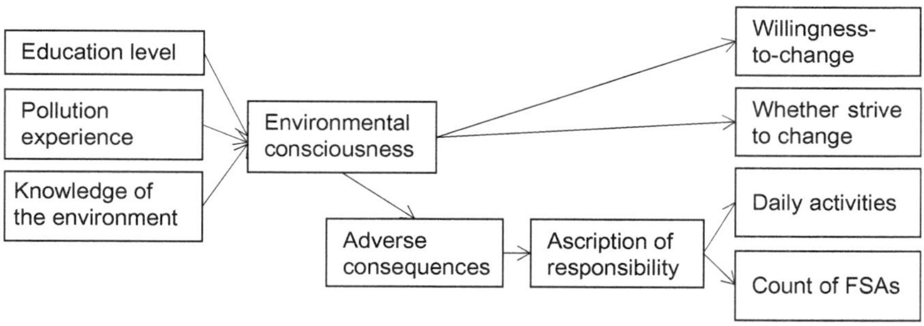

Figure 7.3 Schematic plot of the environmental consciousness and environmental behaviors of the urban residents.

willingness-to-change, whether strive to change for the urban residents. The relationships between the daily activities, the count of FSAs, and the environmental consciousness is stimulated by adding the "adverse consequences" and "sense of obligation" based on the VBN theory. These relationships are pretty similar to rural residents. As a result, the schematic plot of the environmental consciousness and environmental behaviors of the urban residents is also very similar to the rural residents, as shown in Fig. 7.3.

7.4 Environmental consciousness changing and the path to realizing environmental behavior

In the last section, we have combined the findings of the last few chapters and construct the model for each of the resident groups. The migrant group, as discussed in Chapter 4: From a Longitudinal Study to Cross-Sectional Study, allow us to change from longitudinal study to cross-sectional study since it germinated the environmental consciousness from the fixed location and time to the broader dimension of time and locations. In this section, we will compare the three resident groups to find the rule of the changing of the environmental consciousness. Based on it, we will also try to construct the path to realize the environmental behavior for all the three groups of residents.

First, we compare the relationships between the demographic variables and environmental consciousness. By comparing Model-D in Tables 7.1, 7.3, and 7.5, very different explanation power is shown for the three groups of residents. For rural residents, the demographic variables have a high explanatory power on environmental consciousness. In other words, residents' native characteristics have a great influence on their environmental consciousness. Specifically, rural residents who are younger, have a higher education level have a higher level of environmental consciousness. Meanwhile, the demographic variables of urban residents and migrants have a weak influence on their environmental consciousness. Among the demographic variables, only age and education level (only education level) have some influence on the environmental consciousness for the urban residents (migrants), respectively. If we consider the three groups of residents approximated as a continuum from the rural residents to migrants to urban residents, the explanation power for Model-D is monotonically decreasing.

Next, we compare the relationships between the three theoretical dimensions and environmental consciousness. The environmental consciousness of urban residents and migrants is greatly influenced by the theoretical dimensions, as the larger explanation power of Model-T than Model-D shown in Tables 7.1, 7.3, and 7.5. In contrast, the explanation power of Model-T is only around half of Model-D for the rural residents. If we also consider the three groups of residents approximated as a continuum (from the rural residents to migrants to urban residents), the influence from the theoretical dimension is monotonically increasing and the influence from the demographic variables is monotonically decreasing.

Therefore, we believe that the regional differences between urban and rural areas lead to the changing of environmental consciousness. The most important of which is reflected in the impact of demographic variables and theoretical dimensional variables on the environmental consciousness. Through the comparison of the three groups of residents, we believe that the migration from rural to urban areas will make the residents' environmental consciousness more influenced by social factors, and the influence by the demographic variables gradually faded. Factors affecting the level of environmental consciousness are gradually replaced by direct or indirect perceived pollution (such as the impact of constructivism theory). Since China has not yet entered the postmaterialism stage, and the data also verified the weak effect of postmaterial theory on environmental consciousness. As a result, we propose that the change of the environmental consciousness of the three groups of residents can mainly be driven by the pollution-driven theory and constructivism theory.

Moreover, the faith in science and technology has been replaced by the knowledge of the environment from the rural to urban continuum and has become a major factor affecting the environmental consciousness of urban residents. Hong and Xiao (2007) found that for urban residents, environmental knowledge is a mediator between the demographic variables and environmental consciousness. Our research then takes a step forward, that is, we find that the impact of pollution experience on environmental consciousness is very significant. Meanwhile, recognition of the suffering of environmental pollution, that is, pollution awareness, is determined by the education level and the knowledge of the environment (faith in science and technology) for the urban (rural residents and migrants), respectively.

In addition, we can see that the relationship between environmental behaviors and environmental consciousness is more complicated for the migrants and cannot be summarized by any simple existing models. So we try to compare the similarities and differences between migrants and the other two groups. We found an interesting phenomenon, that is, in the complex model of the migrants, some part of the model overlap with the model of the rural residents, and some other part coincides with the model of the urban residents. There are only two variables for the migrants in the model which are not shown for the rural and urban residents: the urban living time, and relative pollution. We then guess that these two variables are crucial for the rural-to-urban environmental consciousness and environmental behavior transformation.

We then perform further analysis of the relationship between environmental consciousness and environmental behavior. We believe that although the environmental behavior and the environmental consciousness are related, they are affected by the situation and have different responses for different depth of actual behavior. The daily activities are not directly related to environmental consciousness, which has been verified in all three groups of residents. In the strive to change behavior, residents mainly choose to discuss with other people in surroundings. In contrast, much fewer residents choose to approach directly to the polluters, approach to a government agency, or report issues to the mass media. What are the factors that relate to daily activities and strive to change behavior, and they are affected by what factors? These are the questions bring out after rule out the direct influence of the environmental consciousness. The VBN theory may provide a model to answer these questions.

In the VBN theory, "adverse consequences" and "ascription of responsibility" are necessary conditions for the transformation of environmental consciousness to environmental behavior. This theory is then tested based on the survey data on the three groups of residents individually.

For rural residents and urban residents, the survey data support the VBN theory. We find that the direct relationship between environmental consciousness and environmental behavior is not significant for the rural resident. After adding "adverse consequences" and "ascription of responsibility," the environmental consciousness is stimulated, and the relationship between environmental consciousness and environmental behavior becomes significant. Explanatory power has also increased. In contrast, for migrants, the survey data does not support the VBN theory. As we discussed in the last section, the environmental behavior of the migrants is affected by both the knowledge of the rules in urban areas and the habitually. In other words, the environmental behavior of the migrants may relate to whether they know the rules in urban areas and whether they use the rules in urban areas to restrain themselves. These two factors may further be considered as "internalization of the new environmental norms" and "sense of belonging to the city" and they bring important connections between environmental consciousness and environmental behavior. In addition, based on the particularity of the migrants and our investigation conclusions, we believe that the ambiguity of "adverse consequences" and "ascription of responsibility" may also imply the verification of the VBN theory. As a result, creating environmental conditions that can stimulate environmental behavior should also become an important part of environmental protection work.

In conclusion, with the attempts of testing the VBN theory, the environmental consciousness and the environmental behavior is analyzed for all the rural residents, urban residents and migrants in China. We then try to construct a different environmental consciousness and environment

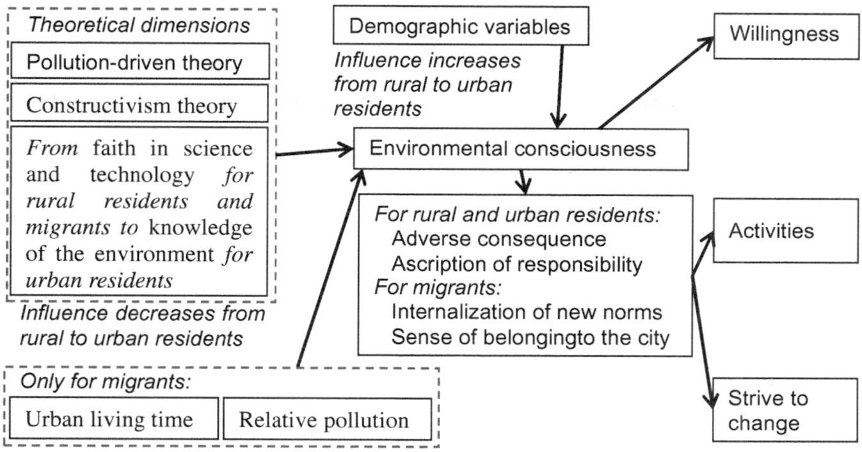

Figure 7.4 Environmental consciousness and environment behavior path map for all the three groups of residents in China.

behavior path map for all the three groups of residents in China, as shown in Fig. 7.4.

7.5 Conclusion and rethinking

In order to reveal the mechanism behind the environmental consciousness and environmental behavior and find the path to environmental behavior, we first analyze the path to the environmental consciousness. We used multivariable linear regression models to analyze the survey data for rural residents, urban residents, and migrants, and compared the explanation power of the models using demographic variables and three theoretical dimensional variables. The stepwise regression is also used to confirm that the variables have influences on the environmental consciousness.

We then analyzed the path from the environmental consciousness to environmental behavior. The VBN theory brings us a good theory model to analyze the data. We have found that the VBN theory works well for the rural and urban resident groups. However, the migrant group of residents still has an unproven relationship between their environmental behavior and environmental consciousness. For the mechanism behind the environmental activities and environmental resistance of the migrants, we believe that the sense of belonging to the city and the internalization of the new environmental norms are the conditions for stimulating pro-environment daily activities and strive to change. These two factors are also related to the urban living time, so we consider to add it in our model for migrants.

After that, we optimized the model of the environmental consciousness and the environmental behavior of the three groups of residents, combined with the demographic variables, the theory dimensions, environmental behavior influencing factors, and the characteristics of each resident group, attempt to systematically construct a model to depict the internal mechanisms of the environmental consciousness and the environmental behavior for the three groups of residents.

Finally, combined with the influencing factors of environmental behavior and environmental consciousness, the practical path to the environmental behaviors of each resident groups are discussed. By consider the three groups of residents in a continuum from the rural residents to migrants to urban residents, the environmental consciousness and environment behavior path map for all the three groups of residents in China is summarized and two findings are listed as follows.

First, among the factors that influence the environmental consciousness, the knowledge of the environment has replaced the faith in science and technology in the process of rural-to-urban transformation and has become a major factor of the environmental consciousness.

Second, the path from the environmental consciousness to environmental behavior implies the extraordinary nature of the migrants. We believe that the significance in the time dimension for the migrants implies the process of "internalization of the new environmental norms" and "sense of belonging to the city."

CHAPTER EIGHT

Environmental consciousness under the tide of the times

Starting from Chapter 3, From Movements to Measurement, we have studied in-depth and meticulously the environmental consciousness and environmental behavior change at the microlevel. Before the end of this book, we hope to bring the picture back to the macrolevel. By using the correlation analysis of the environmental consciousness at both macro- and microlevels, we hope to find a connection between them and help readers to find out the underlying relations of environmental consciousness and environmental behavior.

8.1 Macrolevel impacts

In this section, we will discuss the macrolevel impacts from three aspects, that is, economic impacts, policy impacts, and global interactions.

8.1.1 Economic impacts

One to mirror the environmental consciousness in the United States is the satisfaction of the Government's environmental measures and investments. While the government's decisions depend on economic factors, results can be seen from the comparison. Table 8.1 gives the form of the United States economic situation since the 1950s.

The development of economic factors as well as industrial development in the United States led to the 1970s environmental movement. This movement was based on the effects of increasingly serious environmental problems and the economic strength accumulated in the 1960s in the United States. Both of these factors increased attention to environmental issues and the expanded capacity for the environmental movement. Late in the 1960s and early 1970s, the pillar industry transition and environmental deterioration changed the consideration of the

Environmental Consciousness in China.
DOI: https://doi.org/10.1016/B978-0-08-100388-6.00008-8

Table 8.1 History of the United States economy.

Time	Characteristic	Reason
1950s–1960s	Continued development and prosperity of the western and southern of the United States	The United States is the wealthiest and powerful country of the world. Occupied international markets; promoted science and technology; developed industries and military; improved the level of people living.
1970s–1980s	Decline economic status because of the crisis on 1970s, the economic development be limited	Development took a serious blow by the economic crisis. Inflation and debt became worse.
1990s	Sustainable and stable development in 1990s; economy entered to a new era	Socio-economic reformation; education reinforced; science and technology promotion; information industry development; traditional industries transformation.

environment–economic cost-effective view by the public. And, it also promoted the environmental movement.

From Table 8.1, we can see that there is a strong relationship between the economy and environmental consciousness. Although there is a delayed effect, the period of booming environmental consciousness in the United States is correlated with a prosperous economy. Environmental consciousness can contribute to solving environmental problems. This perspective is similar to the upward Kuznets curve. The Kuznets curve displayed the relationship between economic development and environmental quality as an inverted U-curve. This means when the economy starts developing, environmental deterioration worsened; but, when economic development reached a certain stage, pollution will be controlled, and the environmental quality will gradually be improved.

In addition, the event of 9.11 has had a profound impact on the United States. People began to concentrate on the protection of national security. Bush opposed the Kyoto treaty because of the strain he believed the treaty would put on the economy; he emphasized the uncertainties which he believed were present in the scientific evidence (Barrett, 1998).

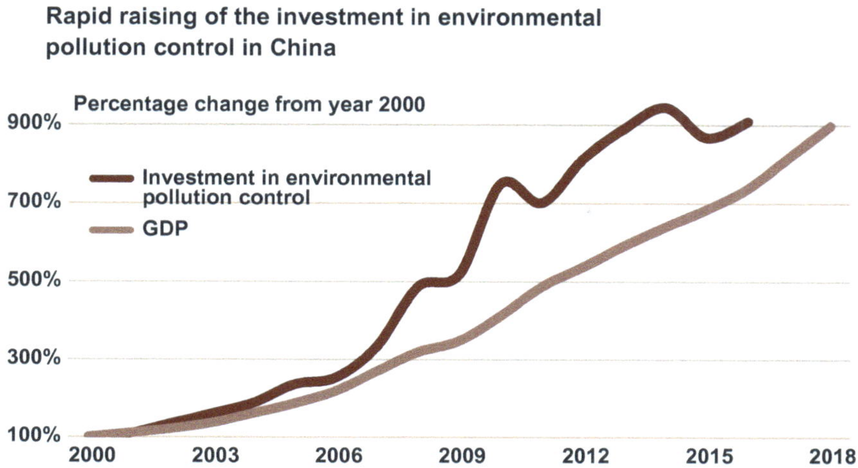

Figure 8.1 The investment in environmental pollution control and GDP in China. *From National Bureau of Statistics of China. (2000–2018).* National Bureau of Statistics of China annual data. <*http://www.stats.gov.cn/english/statisticaldata/AnnualData/*>.

A similar relationship is also found in China. From the year 2000, the economics of China quickly boomed. By plotting the GDP, one of the widely used economic indicator, a quick increase by more than seven times within 16 years is shown in Fig. 8.1. Meanwhile, the Chinese government pays more and more attention to the environmental issues which would lead to a higher and higher environmental consciousness on the macrolevel. To quantify this macrolevel environmental consciousness, we looked for how much the government paid for environmental pollution control which should be a direct environmental behavior resulted from the environmental consciousness at a macrolevel. As shown in Fig. 8.1, the investment in environmental pollution control in China also rapidly increased with a speed higher than the GDP. Within the same time period, the investment (GDP) increased more than nine (seven) times, respectively. This also confirms in China, that the environmental consciousness and environmental behavior in macrolevel is impacted by economics.

8.1.2 Policy impacts

As we saw the economic factors have so many important influences on environmental consciousness. We also can see the role that politics has played in the development of environmental consciousness. From this angle, political influence is mainly manifested in the policy aspect. The policy 42 USC Section 2151p-1, adopted by Congress through P.L. 87-195 December 30, 1969 (National Environmental Policy Act, 42 USC Ch. 55) embarked environmental protection on a legal track. This policy also established the consciousness of environmental protection in the legal system. The major impact on the environment from politics is the laws and regulations. The laws shared similarities in both China and the United States. The examples are the establishment of ecological protection areas, restriction on the production of pollution, and so on. In addition, politics is also associated with the economy and media.

Environmental policies strongly implicated the economy as the "Kyoto Protocol" that was previously mentioned. In addition, industry promotion changed by economic policies in the United States, such as tax adjustment, subsidies increase, and so on. Compared with the macroeconomic background, microeconomic regulation will give more specific control of individuals and businesses. Instead of building an "Economic—Technological development area" as China has done, the government of the United States will use economic policy to lead enterprises.

Policies also have very important influence regarding construction and the impetus of environmental consciousness in China. Chinese environmental protection regulations have continuously improved in the recent years, as is has an influence of environmental consciousness by way of publicity and policy implementation. In other words, the execution of environmental protection in China was passive at the beginning stage. Because environmental consciousness had not been internalized, people's behavior was needed to be controlled by law and stimulated by microeconomic instruments.

In addition, from September 1, 2003 the China Environmental Impact Assessment (EIA) law has become the method and system of analysis, forecasting, and assessment for the project's environmental impacts.

Furthermore, the mass media have also had an important influence on the environmental consciousness of China. The 2008 Olympic Games is an obvious example. To improve the environmental consciousness and promote good performances during the Olympic Games time, the Chinese government started "China Environmental Awareness Program" in June 2006. One of the reports was "2007 China General Public Environmental Survey" which shows that the mass media has become the most important knowledge and information channel for gaining the

Environmental Protection, as seen by the fact that 81.1% of people accepted the environmental knowledge and information through television and radio. People thought environmental issues have become serious social problems in China. In the top 13 ranked issues, environmental pollution was addressed as the fourth, closely after medical needs, employment, and the income gap (Chinese Academy of Social Sciences, 2007).

The influence of politics is not unilateral. The environmental movement in the United States gave us a model of a movement that became a channel for the grassroots aspirations and which in turn was present in presidential campaigns of the United States. Al Gore, Bill Clinton's 1992 campaign partner and the author of Earth in the Balance has shown the environmental desire of people. Earth in the Balance is the 1992 best-selling book which expressed politician willing for environmental concern and determination to protect the earth. As soon as Clinton entered the White House, he gave an "Earth Day" speech which not only showed his support for environmental protection but also inspired the enthusiasm of the "Earth Day."

8.1.3 Global interactions

In the previous subsection, we can see the awakening of the Western environmental consciousness and environmental movement (especially the United States) estimated by the concern of the United Nations (UN). At the same time, the UN concern strengthened environmental consciousness all over the world. The UN reinforced environmental consciousness by giving feedback to the United States and awakened Chinese. Then, although decisions always depend on every situation, policy, the environmental movement, and the United Nations joined as environmental allies.

On June 16, 1972, the Conference on the Human Environment at Stockholm was the UN's first major conference on international environmental issues and marked a turning point in the development of international environmental politics. In 1977, the United Nations Environment Program (UNEP) proposed the concept "eco-development." In 1980, UNEP convened a seminar to discuss the relationship between population, resources, environment, and development. The seminar pointed out that these four factors were closely related to each other and interpromoted to each other. Therefore, the management of the relationship between these four elements is the new development strategy. In 1981, Lester Brown, who gave the first system conception of sustainable development with the book Building a Sustainable Society analyzed a series of economic development issues, such as land desertification, resource depletion, oil shortages, and lack of food; and put forward the channels to control population, protect resource, and exploit renewable resources. In February 1982, Our Common Future put forward the "sustainable development" concept and founded the fundamental approaches for sustainable development. In May 1982, an extraordinary session of the UNEP introduced the "new environment" concept. In 1983, the World Commission on Environment and Development (WCED) was created to address the growing concern "about the accelerating deterioration of the human environment and natural resources and the consequences of that deterioration for economic and social development."

In 1992, the Union of Concerned Scientists (UCS) wrote "World Scientists' Warning to Humanity" with a majority of the Nobel Prize laureates signatures. In the same year, two programmatic documents—"Rio Declaration on Environment and Development" and "Agenda 21"—published on the United Nations Conference on Environment and Development. Realized environmental issues are the major problems faced by human beings and have a close relationship with the development of society, the research of sustainable development has extended from academic to international focus and governments' agenda.

At the end of 1997, more than 150 countries and regions gathered in Kyoto, Japan in order to save the earth from the threat of global warming, by holding the United Nations Framework Convention on Climate Change (UNFCCC or FCCC). The international environmental treaty produced at the United Nations Conference was intended to achieve "stabilization of greenhouse gas concentrations in the atmosphere at a level that would prevent dangerous anthropogenic interference with the climate system." So far in 1999, including China, already 40 countries have joined the "Kyoto Protocol"; as of January 2009, 183 parties have ratified the protocol, which entered into force on February 16, 2005.

China's environmental awareness has developed from the complex path because this land is suffering the consequence of environmental deterioration while occurring the fast economic development, and the domestic and external political pressures. In recent years, environmental protection became a basic national policy in China. Ecological consciousness began to enter the mainstream of macropolicy and institutional reform, and citizens' eco-consciousness has been significantly improved. However, compared with developed countries, the ecological level in

China is still low, and it has greater limitations and imbalance, particularly in rural areas. Due to the socio-demographic variables that could explain much more environmental consciousness in Rural Residents Group than Urban Residents Group, we bold prediction that the aspects of social construction should play a more important part in setting eco-consciousness. Social economy, media literacy, science knowledge, and the chance to participate in ecological activities may be verified in future work.

8.2 Mass media, from macro- to microlevel

In the last few chapters, we have discussed the path from the environmental consciousness and then to environmental behavior. In the new era with information and technology rapidly developing, the mass media started to play a more and more important role in the path, for example, for constructivism and environmental resistance.

From traditional media, like newspapers and television to modern media, like networks, the technological leap of media carriers has brought almost complete coverage for all the three groups of residents. When our survey ended in 2012, the level of environmental consciousness among rural residents was still at a low level, but this low level was quickly broken by our expectations. When rural residents are gradually integrated into the modern media network through new technologies, such as computers and smartphones, their level of environmental consciousness will inevitably increase substantially. This is directly related to the conclusion of our research, that is, constructivism has a large influence on environmental consciousness.

8.2.1 Newspaper

John Hannigan points out "Environmental problems... must be 'constructed' by individuals or organizations who define pollution or some other objective condition as worrisome and seek to do something about it" (Hannigan, 1995). In this book, Hannigan believes that the two major social institutions which have the most impact on the evaluation of environmental events are science and mass media. According to Hannigan's point, science and media play a major role "in constructing environmental risks, knowledge, crises, and solutions" (Hannigan, 1995).

Many researchers also agree with the concept that mass media has a great impact on environmental consciousness in China (Hong, 2005; Song, 2010). Researchers also consider mass media as a major channel for people to get environmental protection knowledge (Yang, Dai, & Dou, 2001). When people judge the environmental issue, they majorly rely on the mass media (China Environmental Awareness Project Office, 2008). Among the mass media, the newspaper is one of the most important and most powerful one. It also plays a major role in contributing to environmental consciousness (China Environmental Awareness Project Office, 2008).

In this research, we selected environment-related news from 31 provinces' newspaper in China. The paper selection is based on two rules. First, only newspapers from capital cities or municipalities could join in the research. We selected 31 typical newspapers in China based on this rule. Second, the news should have the keyword "environmental protection," since based on our reading experience, the news has the keywords "environmental protection" most likely also have other keywords like "environmental issue" and "pollution." However vice versa is not true. So, we choose the keyword "environmental protection." We then applied these two rules on the searches in the China National Knowledge Infrastructure (CNKI). The searches resulting in 8045 news from 2000 to the third quarter of 2019 and Fig. 8.2 shows its changes over time.

Fig. 8.2 shows a clear change during these two decades' time. Before 2005, very few news concerned about environmental protection, that is, only less than a hundred news for each year before 2004 and around 150 news for 2005. This number spurted to more than 500 in the year 2006. Then the number reached a new highest in 2007 and between 2006 and 2011, the number is still more than half a thousand. After 2012, the news related to environmental protection shows some retreat to less than half a thousand till 2016. In 2017, 10 years after the first peak, the number reached its second peak at more than 800. This number remained high in 2018 and 2019 if we compare the monthly number of news for the first three quarters for 2019 to the previous years.

To reveal the relationship between the mass media and the environmental consciousness, we also added a variable we used in the previous section, that is, the investment in environmental pollution control by the government of China. As this variable reflects the environmental consciousness at the macrolevel, the relationship between the two curves in Fig. 8.2 gives the relationship between the environmental consciousness at

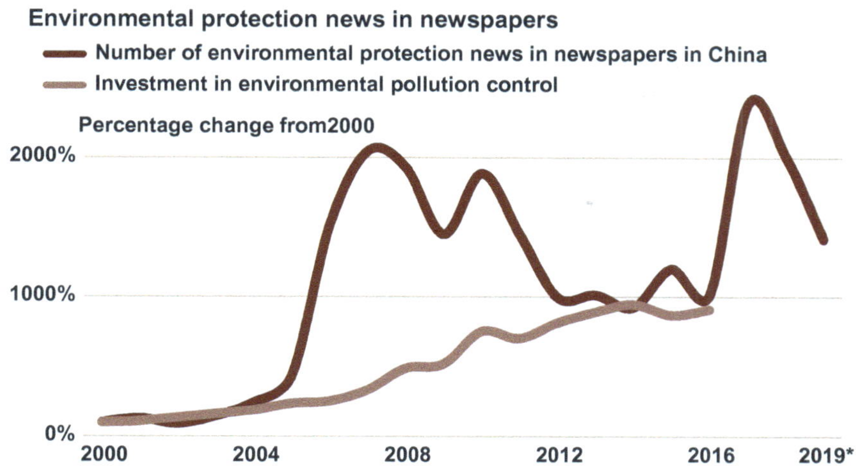

Figure 8.2 Number of news related to environmental protection in newspapers in China and investment in environmental pollution control from 2000 to 2019. *The number of 2019 environmental protection news only counts the first three quarters. *Data source of investment in environmental pollution control from National Bureau of Statistics of China. (2000–2018).* National Bureau of Statistics of China annual data. <*http://www.stats.gov.cn/english/statisticaldata/AnnualData/*>.

the macrolevel and the mass media. Since the mass media reports the major projects by the governments, our hypothesis is the more investment in environmental pollution control, the more major projects relate to it and thus the more news report on these projects or related to environmental protection. By comparing these two curves in Fig. 8.2, we can see they have the same major trends, that is, increases over time. In addition, the number of news curve shows more peak at around 2005–2011. It may imply that some other factors also affect the number of news in newspapers.

8.2.2 Internet

Recent years, our life has been greatly changed by rapidly developed technologies. One of the most important changes is the fast-development of the internet. The history of the internet starts from the 1960s and originated in several computer science laboratories. The first internet connection in China started in 1994, and in the same year, the first web server, "China window," was established in China. In 1995, Beijing had the first advertising board for the internet. It was called an information highway at that time. The first bulletin board system (BBS) in China, Shuimu

Tsinghua, was also established. BBS greatly changed the way we communicate. Sharing and exchanging some thoughts with others became much easier by using the BBS. During the last few years of the 1990s, more and more major websites in China were established like NetEase, Sohu, Tencent, Sina, and Baidu. They provide not only the services of online sharing or exchange of messages like BBS, but also they provide online news, information search, online chats, emails, online games, and many more services. With these services, the way people get information greatly changed. People can search for their interesting topics online instead of only reading the news that the newspapers have. This usually makes the people who are interested in the environment easier to get more knowledge of environments, and thus may increase their environmental consciousness. The easier sharing and exchanging of information may also let the environmental consciousness spread faster.

These changes usually only happen on internet users. The number of internet users also change how much the internet will affect environmental consciousness. As the website servers rapidly increase in China, the number of internet users in China is also increasing rapidly, from around half a million in 1997 to more than 22 million in 2000. As time goes by the year 2006, around 10% of the population in China became internet users. In 2007, Apple announced the first generation of iPhone, which started a new era with a much more convenient way to access the internet. After then, the number of smartphone internet users quickly raised. In 2014, the number of smartphone internet users was more than 80% of all internet users and smartphones became the most widely used platform to access the internet in China. The quick increase of smartphone internet users also brought rapid increase in internet users. In 2015, more than half of the population in China became internet users, and the fraction further increased to more than 60% in 2019. Nowadays, residents not only rely on computers or smartphones to surf on the internet. Tablets, smart TVs, smart home assistants, and even more smart devices are connected online and they make our life more and more convenient. When we enjoy a more convenient life, they also become a more important way for us to get information and affect our environmental consciousness.

Earlier research has found some relationship between the internet and environmental consciousness. Research in Europe found that the internet has a significant role to enhance the environmental consciousness in terms of perceived environmental information, environmental attitudes, and especially environmentally friendly consumerism (Nistor, 2010). In China, another research focused on the web-based environmental groups found that these web-based groups may raise the environmental consciousness (Yang, 2003). In this book, we are trying to view the internet from another aspect. The internet also is a very good tool to study what the users are interested in. An internet user is most likely using search engines to search for the keywords they are interested in. By analyzing the frequency of the keywords, we can find out how the consciousness changes over time. In this book, we use the data from the largest search engine in China, that is, Baidu. Baidu provides a service called Baidu Index which gives how frequently a keyword is searched. Similar to the keyword used for newspapers, we also use "environmental protection" as the keyword and plot the result as shown in Fig. 8.3. Since the Baidu only shows the result from the year 2011, we then choose this year as the starting year for this figure.

Fig. 8.3 shows the change of the Baidu search frequency for the keyword "environmental protection" over the past 9 years. The search frequency almost has no change for the first 6 years, that is, 2011–16. In 2017, the average search frequency increased by more than 20%. We then compared these changes with another mass media we introduced in the last subsection, that is, newspapers. We can see the news related to environmental protections also have a large increase of more than 60% during

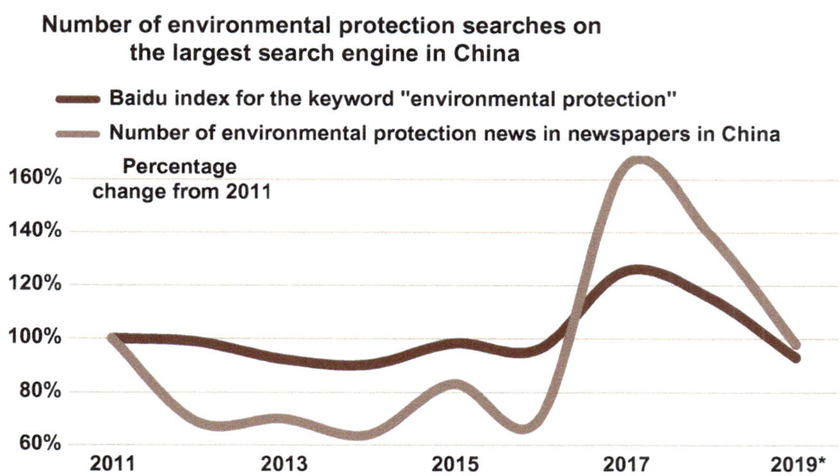

Figure 8.3 Average daily number of searches for the keyword "environmental protection" on Baidu and the number of news related to environmental protection in newspapers in China from 2011 to 2019. *The number of 2019 data only counts the first three quarters. *Data source of average daily number of searches from Baidu Index.*

the same year. For the following 2 years, the average search frequency reduced, and so does the news in newspapers. This result shows these two types of mass media, that is, the traditional newspaper media and the new internet media are significantly correlated. They both play a very important role in the environmental consciousness of the public.

8.2.3 Driven the environmental consciousness

As we reviewed in the earlier part of this section, previous research asserts mass media has a great impact on environmental consciousness. Based on that, our hypothesis is, the public environmental consciousness over time will be correlated to the curve of the mass media's environmental concern.

Since we need to study environmental consciousness, it is a prerequisite to grasping the changing of environmental consciousness over the years. The questionnaire "Ranking popular social issues in China" in the survey for "People's Environmental Protection Livelihood Index in China" is a very good way to reflect environmental consciousness. The "People's Environmental Protection Livelihood Index in China" is the first environmental consciousness index in China organized by the China Environmental Culture Promotion Association and led by the Environmental Protection Administration. Since 1999, it has made continuous investigations over many years till 2008, and is known as the "environmental consciousness barometer of China." From 1999 to 2006, the survey covered 8–28 capital cities/municipalities, then covered all 31 capital cities and municipalities. Therefore, we extract the "environmental concern rate" from this index between 1999 and 2008, as shown in Fig. 8.4.

Although the concern rate was very low in 1999, a sharp arise appeared in 2000, which is the peak of the decade, as shown in Fig. 8.4. The lowest concern rate is shown at the year 2005, then it kept increasing in the following years, with the slim drop at 2008. Does the year changing curve of environmental concern match the mass media's focus? To answer this question, we then compared these two curves. Fig. 8.4 shows the same trend for these two curves from 2005 to 2008. However the trend does not show any relationship from the curves between 2000 and 2005.

This trend leaves a question to the previous hypothesis, how to explain the unmatched trends from 2000 to 2005? This hypothesis was

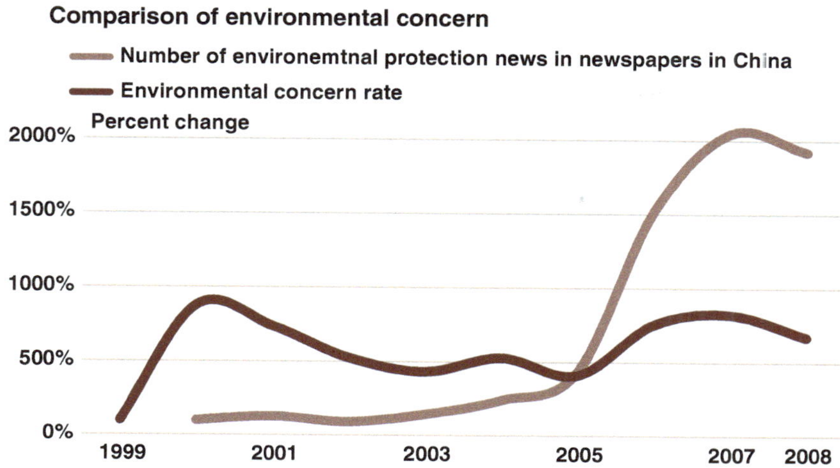

Figure 8.4 Environmental concern rate from People's Environmental Protection Livelihood Index in China and the number of news related to environmental protection in newspapers in China from 1999 to 2008. The percent change is respect to 1999 and 2000 for environmental concern rate and number of news in newspapers, respectively.

built by constructivism theory—the mass media as a channel conduct environmental concern to the public. If the public concern was not directly built by mass media, the reasonable trigger may be the environmental pollution itself, which related to the pollution drive theory. Therefore, our second hypothesis is: the public environmental concern trend will match the environmental pollution changes. According to the "China Environmental Statistics Yearbook," "China Environmental Status Bulletin," "China Environmental Statistics Annual Report," we analyze the number of environmental pollution accidents and the direct economic loss due to the environmental pollution accidents in China between 1999 and 2009.

Generally, the worst year with the maximum number of environmental pollution accidents is 2000. Over two thousand environmental accidents erupted in 2000, and the economic loss was over 170 million yuan, which is over 20 million dollars based on the exchange rate in 2000. We also consider the number of environmental pollution and damage accidents cannot directly reflect the size of pollution incidents. Therefore, this book uses the direct economic loss due to environmental pollution and damage accidents to visually reflect the environmental pollution.

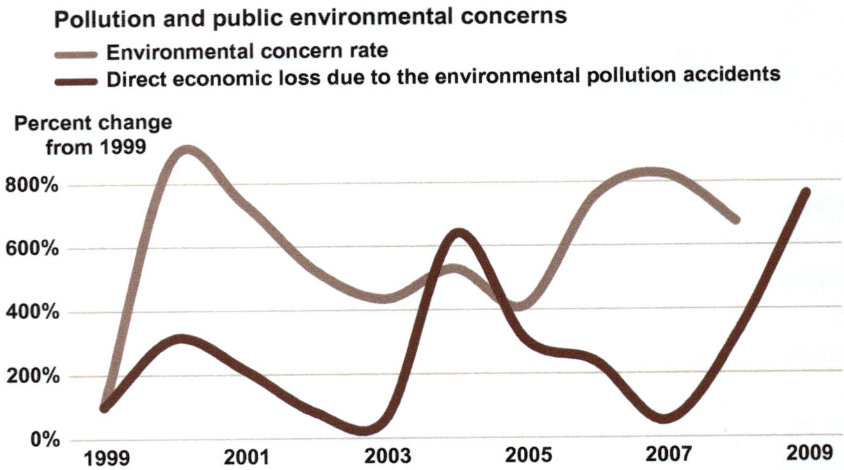

Figure 8.5 Environmental concern rate from People's Environmental Protection Livelihood Index in China and the direct economic loss due to the environmental pollution accidents in China from 1999 to 2009.

Does the curve of environmental pollution direct economic loss match the public environmental concern? We compared these two curves, as shown in Fig. 8.5. We can see by comparing the relationships, that there are two stages, that is, before 2005 or the first stage, and after 2005 or the second stage. In the first stage, the public environmental concern curve and the direct economic loss due to the pollution curve are perfectly correlated. Their peaks and valleys are just consistent in the turning point. The highly matched curves verified our second hypothesis: the public environmental concern is significantly related to environmental pollution.

However, the curve of the second stage since 2005, really gave the big challenge of the result. Especially in 2007, the public environmental concern raised to the peak while the economic loss due to pollution reduced to the bottom. How do we explain this difference? If we take another look at the first hypothesis, that is, the public environmental concerns should relate to the mass media's environmental concern. Although we found the data does not support this hypothesis in general, the data in the second stage did tell a special story in Fig. 8.4. From the figure, we can see they showed the same trend during the second stage. We could have a perfect figure if we select the first stage (1999–2005) of environmental pollution direct economic loss combined with the second stage (2005–08) of mass media's environmental focus to match public environmental concern.

8.3 Summary

To improve public scientific literacy is the necessary requirement of modern societies' civilization. This opinion became the consensus of both the scientific and political arenas. In our daily life, the success of a series of social development issues mostly related to the environment—such as global warming, human cloning, AIDS prevention, and the investment in technology research and so on—are mainly based on the consciousness and participation of people. Therefore, research concerning the establishment of environmental consciousness has helped us in dealing with environmental problems.

Hannigan pointed out that "Environmental problems do not materialize by themselves; rather, they must be 'constructed' by individuals or organizations who define pollution or some other objective condition as worrisome and seek to do something about it" (Hannigan, 1995). This view is related to the mass media "agenda-setting" theory. A well-known dictum by political scientist Bernard Cohen was that "The press may not be successful much of the time in telling people what to think, but it is stunningly successful in telling its readers what to think about" (Cohen, 1963). In the United States, important issues are put on the agenda by mass media; the typical example is environmental protection in the 1970s. Agenda-setting theory shared the same idea with Anthony Downs's "issue-attention cycle"—the rise and fall of an issue on the public agenda (Downs, 1972). However, Riley E. Dunlap points out that "Downs' issue-attention cycle is not an inexorable process" (Dunlap, 1995). Public support for environmental protection has gone through its infancy, arousal, climax, and decline, but is still in sight. Even after the government withdrew its attention from the environment, mass environmental consciousness made the 1980s revive after the recession.

The government indeed received pressure from the environmental movement. On one hand, the environmental movements made the government execute a more real practice while on the other hand, the Green Party grew and gained the opportunity to exercise influence over the government. In order to gain the public's trust and votes, the US's political parties call themselves as environmental protectors.

The Chinese government took responsibility of the society as environment deterioration became more serious. The people related their expectations of environmental protection to the government while the government issued a top-down policy for dealing with environmental issues. In China, environmental protection reflects public reliance on government. It also reflects the government's concern and responsibility for society. The establishment of public environmental consciousness relied on science and technological development as well as the policy publicity through mass media. In addition, particular programs (i.e., the 2008 Beijing Olympic Games) were also used in an agenda-setting manner to raise the level of both environmental consciousness and the action of environmental protection.

The scientific and technological revolution opened the way for modern society, but they also created the possibility of risk to society. For developing countries, it is difficult to cross this gap. They should face the problems that developed countries have experienced, such as rule of law incompletion, economic market imperfection, rapid population grown, labor confliction, sexual discrimination, wealth differentiation, environmental pollution, and so on. At the same time, developing countries faced problems that were not faced by a developed country, such as nationalism and cultural conservatism. In addition, various developed levels in one developing society made it difficult to adapt to change. Furthermore, with the era of globalization, developing countries also have to face the open possibility of risk from developed countries.

Environmental problems are the typical issues include in these topics. Along with the strengthening of globalization, environmental problems have the potential to polymerize or cover the traditional problems; even occupying the position of a major problem. In this way, environmentalism will become more complex and extensive in developing countries (Yan, 2007).

In regards to China's search for solutions to its environmental problems, the experience of Western society is worth learning from. However, China should not give up another road—the traditional culture of life in harmony between nature and humans. Lao Zi, Zhuang Zi, Confucius, and other well-known sages often put "harmony" as the core of though. The Moral sutra said: "Ren Fa Di, Di Fa Tian, Tian Fa Dao, Dao Fa Zi Ran" means all things have to follow the natural law. Master Lü's Spring and Autumn Annals states, "Why not drain water for fish? No fish next year" these traditional verses show how to use natural resources rationally.

From the late 19th century to the mid-20th century, the Chinese lost the numerous outstanding traditional cultures and the principles of moral excellence described by its ancestors. As people are feeling more and more unsafe today, China's new direction may not be the same as the West—following the footsteps of pure technology. But rather, China may find the implicit clues from its own history.

CHAPTER NINE

Future considerations

Environmental stories repeat again and again on the earth. From England to the United States; from Japan to China. A rapid economic boom followed with the environmental issues, this curse is still going around the world.

The reasons of environmental pollution are complex. After environmental pollution, the factors that affect environmental consciousness and environmental behavior are more chaotic. In order to explore the logic behind local environmental consciousness and environmental behavior in China, we analyzed the relationship between environmental pollution, environmental consciousness, and environmental behavior through multiangles and multilevels.

Macrolevel:

1. The book opened with reviewing the history of environmental pollutions. With many background pictures, the book then reviewed the change of human–nature relationships from the ancient era. Environmental awareness/consciousness was introduced by comparing China and the United States' development.
2. The micromeasurement and analysis of Chinese environmental consciousness are also based on macrostructural variables. The authors made an ambitious assumption based on an understanding of China's dualistic social structure. Combination with the urban and rural dual social structure, the inherent accelerator of the rapid rise of environmental consciousness in China is explored.
3. At the end of the book, context variables are introduced to supplement the analysis of China's macrobackground. Based on social changes with the progress of the technology and mass media, the study of China's environmental consciousness and behavior has been embedded in the torrent of historical development.

Microlevel: Group research

As environment deterioration and urbanization engaged in China, the specialty of environment consciousness transformation appears as different groups of people hold different environmental consciousness (and

Environmental Consciousness in China.
DOI: https://doi.org/10.1016/B978-0-08-100388-6.00009-X

environmental behavior). Therefore, this book would like to be an explorer of environmental consciousness research at the microlevel.

The fundamental view of this research is based on the separation of migration groups. Since China is on her way of urbanization, a lot of people move from rural to urban and become migrants. This point of view gives a new insight into environmental consciousness. Most previous research is taking place in urban areas, which may not reflect the consciousness of the overall population. This special insight not only provides a review of previous research but also gives us a chance to predict the change in the future. As urbanization is still in progress, more and more migrants settle down and become urban residents. The transformation of their environmental consciousness could be used to predict future environmental consciousness.

Therefore, in this research, people are divided into urban residents, rural residents, and migrants (from rural to urban). We believed this design would help us understand the whole picture better. The first step of our research confirmed different environmental consciousness levels are observed in the three groups, from urban residents to rural residents, the environmental consciousness decreases. The second step is trying to find the impact factor of environmental consciousness. The third step is to profile the relationship between environmental conciseness and environmental behavior. The main results from this research are listed as follows,

1. The rural resident group holds the lowest environmental conciseness level. In general, younger, higher educated rural residents who believe in science have a higher environmental consciousness than the rural residents who are older, less educated and not believe in science.
2. Urban residents hold the highest environmental consciousness level in the three groups. In general, higher educated urban resident who has more environmental knowledge and has pollution experience will have a higher environmental consciousness.
3. For the migration group, they are holding a much higher environmental conciseness level than rural residents but still lower than urban residents. In general, higher educated migrants who believe in science and technology also suffered pollution and feel stronger about the environmental difference between rural and urban areas, will hold higher environmental conciseness than other migrants.
4. Environmental protection willingness directly related to environmental conciseness, and this relationship applies in all three groups of residents. However, when considering the personal environmental

activities and strive to change, the situation becomes complicated. The relationship between environmental consciousness and environmental behavior can be found when using the value-belief-norm (VBN) theory, that is, adding adverse consequence and ascription of responsibility into the model for the rural and urban resident groups. However, it still not showing significance between environmental consciousness and environmental behavior for migrants even if the VBN theory is considered.

5. The four findings listed above leads two important conclusions: (a) from rural to urban, environmental knowledge replace the faith in science and technology and becomes significant. It is logical as environmental knowledge is part of science which is just more specific to environmental. (b) The relationship between environmental consciousness and environmental behavior mainly reflect the unique properties of the migrants. We believe that the significates of the migrants in the time dimension imply the process of internalization of new norms and a sense of belonging to the city.

Thirty-years of economic reform and opening-up have promoted China's growth and prosperity. However, just like the other big economic bodies in history, China has not escaped the economic–environmental problem loops.

China pays a huge environmental price for the fruits of economic development and this arouses the change. China made commitments to protect the environment and processed top to bottom change including: set up environmental laws and regulations; subsidy environmental-friendly products; advocate public green-consumption; and reduce the impact of economic development on the environment. However, the relationship between the natural environment and the human living environment is more intricate and complex. It will be a long way to go finding solutions to environmental problems for China and for human beings as a whole.

The awakening process from the bottom up is ongoing. Chinese environmental consciousness at the microlevel coincides with the government's goal of environmentally sustainable development. Based on our research findings and analysis, we expect that the overall level of public environmental consciousness in China will continue to rise in the future, although it may not be as rapid as the previous decade. Internal migration will gradually progress in a stable future, and residents' environmental consciousness level will also become a higher and stable level, as long as the environmental behavior. According to our model expectation, in the future, China will be benefited by strengthening environmental education for rural and migrant groups; assisting them to integrate into the urban living environment; stimulating their environmental protection enthusiasm; helping them build ownership.

At the same time, China is playing an increasingly important role in the global environment. Although in the process of policy adjustment, short-term discord may arise among countries. But we believe that protecting the global environment is protecting our own home, and this is the responsibility of every citizen in the world. The ultimate goal is not different.

Earth, home.

References

Ajzen, I. (1985). From intentions to actions: A theory of planned behavior. In J. Kuhl, & J. Beckman (Eds.), *Action-control: From cognition to behavior* (pp. 11–39). Heidelberg: Springer.

Ajzen, I. (2002). Perceived behavioral control, self-efficacy, locus of control, and the theory of planned behavior. *Journal of Applied Social Psychology, 32*, 665–683.

Arcury, T. A., Scollay, S., & Johnson, T. P. (1987). Sex differences in environmental concern and knowledge. *Sex Roles, 16*, 463.

Barrett, S. (1998). Political economy of the Kyoto protocol. *Oxford Review of Economic Policy, 14*(4), 20–39.

Baxter, R. (1988). *Gender differences in attitudes toward nuclear power: A multivariate explanation*. Pullman: Unpublished Doctoral Dissertation, Washington State University.

Beijing Environmental Monitoring Center. (2018). *Beijing environmental monitoring center* (Beijing Environmental Monitoring Center). Retrieved from http://www.bjshjjc.com/index.asp. Last accessed 15.12.19.

Bell, M. L., & Davis, D. L. (2001). Reassessment of the lethal London fog of 1952: Novel indicators of acute and chronic consequences of acute exposure to air pollution. *Environmental Health Perspectives, 109*, 389.

Bird, D. (1967). *November Smog Killed 168 Here (The New York Times)*. Retrieved from https://timesmachine.nytimes.com/timesmachine/1967/10/27/82162296.html?pageNumber = 56. Last accessed 15.12.19.

Blowers, P. G. (1995). *Environmental policy in an international context*. Oxford, UK: Butterworth-Heinemann.

Boldero, J. (1995). The prediction of household recycling of newspapers: The role of attitudes, intentions, and situational factors. *Journal of Applied Social Psychology, 25*(5), 440–462.

Brimblecombe, P. (1987). *The bog smoke: A history of air pollution in London since medieval times*. Abingdon, UK: Routledge.

Butcher, G. (2013, April). *In the fog about smog: Solving the smog puzzle on earth and from space* (American Chemical Society). Retrieved from https://www.acs.org/content/acs/en/education/resources/highschool/chemmatters/past-issues/archive-2012-2013/in-the-fog-about-smog.html. Last accessed 15.12.19.

Buttel, F. H., & Flinn, W. L. (1974). The structure of support for the environmental movement. *Rural Sociology, 39*(1), 56–69.

Cai, Y. (1989). Evolution of the ideas of human-environment relationship. *Studies in Dialectics of Nature*, 48–53.

Cao, B. (2011). *Migration and social harmony symposium*. Jiangsu Federation of Philosophy and Social Science.

Carson, R. (1962). *Silent spring*. Boston: Houghton Mifflin.

Carton, A.D. (2010). *Environmental worldview and faith in science as moderators of the relationship between beliefs about and attitudes toward nuclear energy*. Atlanta: Psychology Theses, Georgia State University.

Catton, W. R., & Dunlap, R. E. (1978). Environmental sociology: A new paradigm. *The American Sociologist, 13*(1), 41.

Chai, J. (2015). 柴静雾霾调查：穹顶之下同呼吸共命运.

China Environmental Awareness Project Office. (2008). 2007 national environmental public awareness survey. *World Environment* (2), 72–77.

China Environmental Culture Promotion Association. (2005–2009). *Chinese public environmental protection people's livelihood index*. The main content of the survey is the public's perception and attention to environmental issues, which may not directly compare environmental consciousness. However, the survey shows a year-on-year increase in the environmental consciousness in China.

China's Public Environmental Literacy Assessment Indicators Research Project Team. (2010). *China's public environmental literacy assessment indicators research*.

Chinese Academy of Social Sciences. (2007). *China general public environmental survey*. China Environmental Awareness Program.

Chou, L., Huang, C., & Zhou, Z. (2007). 古代西安地下水污染及其对城市发展的影响. 西北大学学报: 自然科学版, *2*, 326.

Cohen, B. C. (1963). *The press and foreign policy*. Princeton, NJ: Princeton University Press.

Cordano, M., Welcomer, S. A., & Scherer, R. F. (2003). An analysis of the predictive validity of the new ecological paradigm scale. *The Journal of Environmental Education, 34* (3), 22.

Davidson, D. J., & Freudenburg, W. R. (1996). Gender and environmental risk concerns: A review and analysis of available research. *Environment and Behavior, 28*(3), 302.

Downs, A. (1972). Up and down with ecology—the "issue-attention cycle". *Public Interest, 28*, 38.

Dunlap, R. E. (1995). Public opinion and environmental policy. In J. P. Lester (Ed.), *Environmental politics and policy* (2nd ed., pp. 63–114). Durham: Duke University Press.

Dunlap, R. E., & Jones, R. E. (2002). Environmental concern: Conceptual and measurement issues. In R. E. Dunlap, & W. Michelson (Eds.), *Handbook of environmental sociology*. Westport, CT: Greenwood Press.

Dunlap, R. E., & Marshall, B. K. (2007). "Environmental Sociology." In Clifton D. Bryant & Dennis L. Peck (Eds.) *21st Century Sociology: A Reference Handbook* (pp. 329–340), Vol. 2. Thousands Oaks, CA: Sage.

Dunlap, R. E., & van Liere, K. D. (1978). The "new environmental paradigm". *The Journal of Environmental Education, 9*(4), 10.

Dunlap, R., & van Liere, K. D. (1984). Commitment to the dominant social paradigm and concern for environmental quality. *Social Science Quarterly, 65*, 1013.

Dunlap, R. E., Liere, K. D., Mertig, A. G., & Jones, R. E. (2000). New trends in measuring environmental attitudes: Measuring endorsement of the new ecological paradigm: A revised NEP scale. *Journal of Social Issues, 56*(3), 425.

Dunlap, R. E., & Mertig, A. G. (1992). *American Environmentalism: The US Environmental Movement 1970–1990*. Philadelphia, PA: Taylor & Francis.

Dunlap, R. E., & Scarce, R. (1991). The polls-poll trends environmental problems and protection. *Public Opinion Quarterly, 55*, 651.

Eagly, A. H., & Kulesa, P. (1997). Attitudes, attitude structure, and resistance to change. In M. H. Bazerman (Ed.), *Environment, ethics, and behavior: The psychology of environmental valuation and degradation* (p. 122). San Francisco, CA: New Lexington Press.

Ekins, P. (1993). "Limits to growth" and "sustainable development": Grappling with ecological realities. *Ecological Economics, 8*(3), 269.

Environmental Health Department. (n.d.). *Minamata disease the history and measures, Chapter 2*. Retrieved from Ministry of the Environment, Government of Japan: http://www.env.go.jp/en/chemi/hs/minamata2002/ch2.html. Last accessed 15.12.19.

Foster, J. B. (2002). *Ecology Against Capitalism*. New York, NY: Monthly Review Press.

Fransson, N., & Gärling, T. (1999). Environmental concern: Conceptual definitions, measurement methods, and research findings. *Journal of Environmental Psychology, 19*(4), 369.

Freese, B. (2004). *Coal: A human history*. New York, NY: Penguin Books.

Gao, C., & Meng, X. (2011). 提高公众环境意识实证分析. Academic Frontier of People's Forum. Retrieved from http://www.rmlt.cn/qikan/2011-04-25/20589.html. Last accessed 15.12.19.

Gardner, G., & Stern, P. (2002). *Environmental problems and human behavior* (2nd ed.). Boston, MA: Pearson Custom Publishing.

Gatersleben, B., Steg, L., & Vlek, C. (2002). Measurement and determinants of environmentally significant consumer behavior. *Environment and Behavior*, *34*(3), 335.

Glasbergen, P., & Blowers, A. (1995). *Environmental policy in an international context: perspectives*. Oxford, UK: Butterworth-Heinemann.

Guagnano, G. A., Stern, P. C., & Dietz, T. (1995). Influences on attitude-behavior relationships. A natural experiment with curbside recycling. *Environment and Behavior*, *27* (4), 699–718.

Hannigan, J. A. (1995). *Environmental sociology: A social construction perspective*. London: Routledge.

Hansla, A., Gamble, A., Juliusson, A., & Gärling, T. (2008). Psychological determinants of attitude towards and willingness to pay for green electricity. *Energy Policy*, *36*(2), 768.

Hayes, B. C. (2001). Gender, scientific knowledge, and attitudes toward the environment. *Political Research Quarterly*, *54*, 657.

He, Q., Jing, W., & Wang, Y. (2004). *Introduction to environmental science*. Beijing: Tsinghua University Press.

He, Y., & Lang, H. (2009). 植被建设在黄土高原水土保持中的意义及其对策. 水土保持研究, *16*, 30.

Hines, J. M., Hungerford, H. R., & Tomera, A. N. (1986). Analysis and synthesis of research on responsible environmental behavior: A meta analysis. *Journal of Environmental Education*, *18*(2), 1.

Hong, D. (1999). Western Environmental Sociology Research. 社会学研究, *2*, 83.

Hong, D. (2001). *Social changes and environmental issues: A sociological interpretation of environmental issues in contemporary China*. 首都师范大学出版社.

Hong, D. (2005). The environmental consciousness of urban residents in China. *Jiangsu Social Sciences*, *1*, 127.

Hong, D., & Lu, C. (2011). Multilevel analysis of public environmental concern: Based on 2003 CGSS dataset. *Sociological studies*, *6*, 154.

Hong, D., & Xiao, C. (2006). Measurement of environmental concern: Application of the NEP scale in China. *Social Science Journal*, *26*(5), 71.

Hong, D., & Xiao, C. (2007). Sociological analysis on gender difference of environmental concern. *Sociological studies*, *2*, 111.

Hu, J. (2005). The Communist Party of China Central Committee's Proposal on the eleventh five-year program on national economy and social development.

Hughes, J. D. (2014). *Environmental problems of the Greeks and Romans*. Baltimore, MD: Johns Hopkins University Press.

Hunter, L., & Rinner, L. (2004). The association between environmental perspective and knowledge and concern with species diversity. *Society and Natural Resources*, *17*(6), 517.

Inglehart, R. (1977). *The silent revolution: Changing values and political styles among western publics*. Princeton, NJ: Princeton University Press.

Javers, E. (2009, May 18). *Obama announces new fuel standards*. Retrieved from Politico: https://www.politico.com/story/2009/05/obama-announces-new-fuel-standards-022650. Last accessed 15.12.19

Jones, J.M. (2016, Apr 22). *Americans' identification as "environmentalists" down to 42%*. Retrieved from Gallup: https://news.gallup.com/poll/190916/americans-identification-environmentalists-down.aspx. Last accessed 15.12.19.

Kaiser, F. G., Wlfing, S., & Fuhrer, U. (1999). Environmental attitude and ecological behavior. *Journal of Environmental Psychology*, *19*, 1.

Kang, K. H., Stein, L., Heo, C. Y., & Lee, S. (2012). Consumers' willingness to pay for green initiatives of the hotel industry. *International Journal of Hospitality Management*, *31* (2), 564.

Kok, G., & Siero, S. (1985). Tin recycling: Awareness, comprehension, attitude, intention and behavior. *Journal of Economic Psychology*, *6*(2), 157–173.

Kong, L. (2015, 12 1). 京城雾霾史上罕见 浓度逼近*1952*年伦敦烟雾事件. (财新) Retrieved from http://china.caixin.com/2015-12-01/100880241.html. Last accessed 15.12.19.

Kotchen, M. J., & Reiling, S. D. (2000). Environmental attitudes, motivations, and contingent valuation of nonuse values: a case study involving endangered species. *Ecological Economics*, *32*(1), 93.

Lan, H. (2015, April 16). *Heavy sandstorm shrouds Beijing* (Xinhuanet). Retrieved from http://en.people.cn/n/2015/0416/c90882-8878833.html. Last accessed 15.12.19.

Li, J. (2000). 人口与环境的理论研究及其意义. 市场与人口分析, *6*(3), 50.

Liu, Z. (2015, December 1). 北京持续重度雾霾 网民："北京没了" (BBC). Retrieved from https://www.bbc.com/zhongwen/simp/china/2015/12/151201_beijing_air_pollution. Last accessed 15.12.19.

Lu, C., & Hong, D. (2011). 建构环境关心的测量模型 基于2003中国综合社会调查数据. 社会, *1*, 35.

Lu, T. (2004). 环境社会学研究综述——对环境社会学学科定位问题的讨论. 社会学研究, *4*, 8.

Maddison, A. (2008). *Angus Maddison's homepage* (Angus Maddison). Retrieved from http://www.ggdc.net/maddison/oriindex.htm. Last accessed 15.12.19.

Maloney, M. P., & Ward, M. P. (1973). Ecology: Let's hear from the people: An objective scale for the measurement of ecological attitudes and knowledge. *American Psychologist*, *28*(7), 583.

McEvoy, J. (1972). The American concern with the environment. In W. R. Burch, N. H. Cheek, & L. Taylor (Eds.), Social behavior, natural resources, and the environment *(p. 214)*. New York, NY: Harper & Row.

Ministry of Ecology and Environment of the People's Republic of China. (1989, December 26). 中华人民共和国环境保护法. Retrieved from Ministry of Ecology and Environment of the People's Republic of China: http://zfs.mee.gov.cn/fl/198912/t19891226_75912.htm. Last accessed 15.12.19.

Miou, S. (2015, April 15). *Sky turns yellow as heavy sandstorm shrouds N. China* (Xinhuanet). Retrieved from http://www.xinhuanet.com/english/2015-04/15/c_134154430.htm. Last accessed 15.12.19.

Mitchell, R. C., & Davies, J. C. (1978). *The United States environmental movement and its political context: An overview*. Washington, DC: Resources for the Future.

Mohai, P. (1992). Men, women, and the environment: An examination of the gender gap in environmental concern and activism. *Society and Natural Resources*, *5*(1), 1.

Morrison, W.M. (2013). *China's economic rise: History, trends, challenges, and implications for the United States*. Congressional Research Service.

National Bureau of Statistics of China. (2000-2018). *National Bureau of Statistics of China annual data*. Retrieved from http://www.stats.gov.cn/english/statisticaldata/AnnualData/. Last accessed 15.12.19.

National Bureau of Statistics of China. (2010). *The Sixth National Population Census Data Bulletin*.

Nemery, B., Hoet, P., & Nemmar, A. (2001). The Meuse Valley fog of 1930: An air pollution disaster. *The Lancet*, *357*(9257), 704–708.

Nistor, L. (2010). The role of the Internet in shaping environmental concern. A focus on post-communist Europe. *Journal of Comparative Research in Anthropology and Sociology, 1*, 145–167.

Nixon, R.M. (1970, July 9). Special Message from the President to the Congress about Reorganization Plans to Establish the Environmental Protection Agency and the National Oceanic and Atmospheric Administration. Retrieved from http://www.epa.gov/history/org/origins/reorg.htm. Last accessed 15.12.19.

Obama, B. (2009). *American Recovery and Reinvestment Act of 2009*. Retrieved from U.S. Goverment Publishing Office: https://www.gpo.gov/fdsys/pkg/BILLS-111hr1enr/pdf/BILLS-111hr1enr.pdf. Last accessed 15.12.19.

Organization for Economic Co-operation and Development. (2007). *OECD environmental performance reviews*. China: OECD Publications.

Pan, Y. (2005). China's environmental issues are mainly caused by our distorted concept of development. *Fortune Global Forum*.

Passino, E. M., & Lounsbury, J. W. (1976). Sex differences in oppositionto and support for construction of a proposed nuclear power plant. In P. Suedfeld, & J. A. Russell (Eds.), *The behavioral basis of design* (p. 1). Stroudsburg, PA: Dowden, Hutchinson & Ross.

Poortinga, W., Steg, L., & Vlek, C. (2004). Values, environmental concern, and environmental behavior: a study into household energy use. *Environment and Behavior, 36*(1), 70.

Population censuses. (n.d.). *Population censuses, 1% population surveys, and family statistics*.

Qu, G. (2000). *Dream and look forward: Past and future of environmental protection in China*. Beijing: China Environmental Science Press.

Qu, G. (2002). *The path of development from Stockholm to Johannesburg* (Vol. 12). Chinese Environmental Protection Industry.

Ren, L. (2002). 环境保护中的公众参与. In M. Yang (Ed.), 环境问题与环境意识. Beijing: 华夏出版社.

Ruttan, V. (1993). Population growth, environmental change, and innovation: Implications for sustainable growth in agriculture. In C. J. Torrey (Ed.), *Population and land use in developing countries* (pp. 124–126). Washington, DC: National Academy Press.

Sancton, T.A. (1989, January 2). Planet of the year: What on earth are we doing? *Time Magazine*.

Sawitri, D. R., Hadiyanto, H., & Hadi, S. P. (2015). Pro-environmental behavior from a socialcognitive theory perspective. *Procedia Environmental Sciences, 23*, 27–33.

Schuhwerk, M. E., & Lefkoff-Hagius, R. (1995). Green or non-green? Does type of appeal matter when advertising a green product? *Journal of Advertising, 24*, 45–54.

Scott, D., & Willits, F. K. (1994). Environmental attitudes and behavior: A Pennsylvania survey. *Environment and behavior, 26*(2), 239.

Seinfeld, J. H. (2004). Air pollution: A half century of progress. *AIChE Journal, 50*(6), 1096–1108.

Shanghai Night Bridge View. (2014). Retrieved from http://www.tripdv.com/news/. Last accessed 15.12.19.

Shen, J., & Saijo, T. (2008). Reexamining the relations between socio-demographic characteristics and individual environmental concern: Evidence from Shanghai data. *Journal of Environmental Psychology, 28*(1), 42.

Sierra Club. (2012). *History: Sierra Club Timeline*. Retrieved from Sierra Club: http://vault.sierraclub.org/history/timeline.aspx. Last accessed 15.12.19.

Song, Y. (2010). 发达地区农民环境意识调查分析——以苏州市714个样本为例. *Chinese Rural Economy, 1*, 53–62.

State Council of the People's Republic of China. (2014). 国务院关于调整城市规模划分标准的通知. 国发 [2014] 51号. Retrieved from http://www.gov.cn/zhengce/content/2014-11/20/content_9225.htm. Last accessed 15.12.19.

Stern, P. C. (2000). Toward a coherent theory of environmentally significant behavior. *Journal of Social Issues*, *56*(3), 407.

Stern, P. C., Dietz, T., Abel, T., Guagnano, G. A., & Kalof, L. (1999). A value-belief-norm theory of support for social movements: The case of environmentalism. *Research in Human Ecology*, *6*(2), 81–97.

Stern, P. C., Dietz, T., & Guagnano, G. A. (1995). The new ecological paradigm in social-psychological context. *Environment and Behavior*, *27*, 723.

Stern, P., Dietz, T., & Karlof, L. (1993). Values orientation, gender, and environmental concern. *Environment and Behavior*, *25*(3), 322.

Stevens, J. P. (1999). *Intermediate statistics: A modern approach*. Hillsdale, NJ: Lawrence Erlbaum Associates.

Sui-Lee Wee, B.B. (2015, 12 7). *Still air traps smog in China's capital triggering 'red alert'* (Reuters). Retrieved from https://www.reuters.com/article/us-china-pollution-idUSKBN0TR0C920151208. Last accessed 15.12.19.

Sun, Y., & Wu, C. (2007). Review on the development of environmental behavior theory. *Science Research Management*, *28*(3), 108.

Taylor, S., & Todd, P. (1995). An integrated model of waste management behavior: A test of household recycling and composting intentions. *Environment and Behavior*, *27* (5), 603–630.

The Leading Group Office of Migrant Workers in Jiangsu Province. (2007). *Investigation report of migrant workers in Jiangsu Province*. Nanjing, China: Southeast University Press.

The Official Website of the Beijing 2008 Olympic Games. (2008). *The official website of the Beijing 2008 Olympic Games*. Retrieved from The official website of the Beijing 2008 Olympic Games: https://www.olympic.org/beijing-2008. Last accessed 15.12.19.

Tindall, D. B., Davies, S., & Mauboules, C. (2003). Activism and conservation behavior in an environmental movement: The contradictory effects of gender. *Society & Natural Resources*, *16*(10), 909.

Tong, Y. (2002). Environmental awareness and orientation of environmental policy. In M. Yang (Ed.), *Environment issues and environmental awareness* (p. 65). Beijing: Huaxia Publishing House.

Townsend, J. G. (1950). Investigation of the smog, incident in Donora, Pa., and vicinity. *American Journal of Public Health*, *40*, 183.

Tracton, S. (2012, 12 20). *The killer London smog event of December, 1952: A reminder of deadly smog events in U.S.* (The Washington Post). Retrieved from https://www.washingtonpost.com/blogs/capital-weather-gang/post/the-killer-london-smog-event-of-december-1952-a-reminder-of-deadly-smog-events-in-us/2012/12/19/452c66bc-498e-11e2-b6f0-e851e741d196_blog.html. Last accessed 15.12.19.

UCLA Library. (1948). *View of part of the Los Angeles Civic Center masked by smog in 1948*. Retrieved from Los Angeles Times Photographic Archive, UCLA Library: http://digital2.library.ucla.edu/viewItem.do?ark = 21198/zz0002tsmb&imageonly = true. Last accessed 15.12.19.

U.S. Department of State Air Quality Monitoring Program. (2018). *Beijing air quality historical data* (U.S. Department of State Air Quality Monitoring Program). Retrieved from http://www.stateair.net/web/historical/1/1.html. Last accessed 15.12.19.

van Liere, K., & Dunlap, R. E. (1980). The social bases of environmental concern: A review of hypotheses, explanations and empirical evidence. *Public Opinion Quarterly*, *44*(2), 181.

Vlosky, R. P., Ozanne, L. K., & Fontenot, R. J. (1999). A conceptual model of US consumer willingness-to-pay for environmentally certified wood products. *Journal of Consumer Marketing*, *16*(2), 122–140.

Wang, Z. (1955). *The relationship of geographical environment, population and social development*. Shanghai: The New Knowledge Press.

Wei, L. (2013, 11 21). 改革开放中国经济发展成就：33年GDP年均增长9. 8%. (中国网). Retrieved from http://guoqing.china.com.cn/2013-11/21/content_30659371.htm. Last accessed 15.12.19.

Weigel, R., & Weigel, J. (1978). Environmental concern, the development of a measure. *Environment and Behavior*, *10*(1), 3.

World Bank. (2009). China from poor areas to poor people: China's evolving poverty reduction agenda - An assessment of poverty and inequality in China. The World Bank.

World Health Organization. (1991). *WHO meeting on the assessment of quality of life in health care*. Geneva: WHO. (MNH/PSF/91.4).

Wu, Z. (1950). Geographical environment and social development. Shanghai: Shanghai Tangli Press.

Xinhua Daily Telecommunications. (2016, December 2). 改革开放30多年, 中国逾7亿人脱贫. (Xinhua net). Retrieved from http://www.xinhuanet.com/mrdx/2016-12/02/c_135875062.htm. Last accessed 15.12.19.

Xu, G. (2006). 西方环境社会学研究的三种范式. 中国环境管理干部学院学报, *2*, 24.

Xu, Q. (1997). *Survey of public environmental awareness in China*. Beijing: China Environment Science Press.

Yan, B. (2008). 社会变迁中的环境法. Wuhan: Huazhong University of Science and Technology Press.

Yan, G., Kang, J., Xie, X., Wang, G., Zhang, J., & Zhu, W. (2010). Change Trend of Public Environmental Awareness in China. *China Population Resources And Environment*, 20(10), 55.

Yan, M. (2007, November 9). 绿色思潮在中国兴起. Retrieved from Sociology prespective: http://www.sociologyol.org/shehuibankuai/shehuiredian/shehuizhuyishengtaiwenming/2007-11-09/3920.html. Last accessed 15.12.19.

Yang, G. (2003). Weaving a green web: The internet and environmental activism in China. *China Environment Series*, *6*, 89–93.

Yang, G. (2008). 探究教师的可持续发展教育信念——发展教师信念量表的考虑. 外国教育研究, *9*, 5.

Yang, L., Dai, M., & Dou, Y. (2001). On the constitutes, structure and development of environmental consciousness. *China Environmental Science*, *6*, 545–548.

Yang, S. (2015, 4 16). 北京沙尘来袭呼唤环保“公众参与”提速. (人民网). Retrieved from http://env.people.com.cn/n/2015/0416/c1010-26853401.html. Last accessed 15.12.19.

Zelezny, L. C., Chua, P. P., & Aldrich, C. (2000). New ways of thinking about environmentalism: elaborating gender differences in environmentalism. *Journal of Social Issues*, *56*(3), 443.

Zeng, P. (2005). *China in the world economy*. Beijing: China Development Forum.

Zha, X. (2008, August 18). 北京奥运的蓝天行动：与2. 5微米颗粒战斗. Retrieved from 人民网: http://env1.people.com.cn/GB/7681874.html. Last accessed 15.12.19.

Zhong, Y., Tan, Q., & Zhang, Y. (2003). 大学生环境意识与环境行为的调查研究. *Psychological Science*, *3*, 542.

Zhou, S. (2009). *Scientific concept implementation and environmental protection development*. Seeking Truth Vol. 5.

Zhou, S. (2011). *Environment protection system needs support from technology*. China Environmental Protection Industry, p. 4–10.

Zhou, Z. (2008). Environmental awareness research: State, crisis and solutions. *Journal of Xiamen University (Arts & Social Sciences)*, *4*, 19.

Zhu, L., & Chen, R. (2004). *New migrants in city--Research report of Nanjing*. Nanjing: Nanjing University Press.

Index

Note: Page numbers followed by "*f*" and "*t*" refer to figures and tables, respectively.

FSC
www.fsc.org
MIX
Papier | Fördert
gute Waldnutzung
FSC® C083411

Kolibri 360 GmbH
Ferdinand-Jühlke-Straße 7
99095 Erfurt, Deutschland
produktsicherheit@kolibri360.de